From the very earliest days of Operation Barbarossa, the German invasion of the Soviet Union, it was apparent that the Wehrmacht's standard 3.7cm and 5.0cm anti-tank guns were at a decided disadvantage when faced with the latest Russian armoured vehicles. Increasingly, the Pzkpfw IV with its 7.5cm L/24 weapon was pressed into service as a tank killer but it had in fact been designed as a bunker buster, intended to support infantry assaults, and was allocated to the Panzer regiments in limited numbers. Only the 8.8cm guns of the Flak battalions, which had been used successfully in an anti-tank role in France and North Africa, could be sure of stopping a monster such as the KV-2.

The expedients of early 1942 were based, in the main, on the chassis of obsolete tanks and in many cases employed captured weapons, for example the Russian 7.62cm anti-tank gun. These vehicles proved to be effective against the Red Army's T-34 and KV types but it was obvious that a more powerful gun with a longer range would soon be needed. In addition, these Panzerjäger, or tank hunters, were lightly armoured and the crews were particularly vulnerable to airburst rounds or even determined infantrymen.

The development of an armoured vehicle that incorporated a fully-enclosed fighting compartment with a large-calibre weapon, capable of outranging the enemy's tank and anti-tank guns, was underway but it would be some time before these plans could be expected to come to fruition (1).

In June 1942, at a meeting to discuss armament development, Hitler's attention was drawn to the 8.8cm Flak 41 which was just then entering production. Informed of the gun's range and performance, the Führer immediately ordered that an anti-tank version be built. Within weeks he demanded that work on the new gun be pushed forward and at some time in September 1942 it was decided that both a towed version and a self-propelled gun would be manufactured.

At that time the firm of Altmärkische Kettenwerke (Alkett) of Berlin was engaged in building a chassis and superstructure which could accommodate the 15cm sFH 18 heavy howitzer and it was decided that versions equipped with the howitzer and the 8.8cm gun would be produced. Both types were presented for Hitler's inspection on 2 October 1942 and he ordered that 100 of each variant be available by mid-May 1943.

Initially christened Hornisse, and later renamed Nashorn, the self-propelled anti-tank guns were assigned to newly-formed battalions that would operate under army-level commands, as would the towed version, and the bulk were sent to the Eastern Front (2).

Less than 500 examples of the Nashorn were assembled from February 1943 to March 1945 (3) and it was always regarded as a transitional solution, bridging the gap between the improvised Panzerjäger and the purpose-built Jagdpanzer. Nevertheless, its main gun was a fearsome weapon and although some post-war claims as to its efficacy should be viewed with circumspection, official after-action reports consistently suggested that the gun's optimum range was 2,000 metres. The Nashorn's failings came to the fore when it was employed as a tank, which was actually forbidden, but in its intended role as an anti-tank weapon, firing from concealed positions, it had few equals.

Notes

1. The history of these vehicles is examined in *TankCraft 8: Jagdpanther Tank Destroyer, German Army and Waffen-SS, Western Europe 1944-1945* and *TankCraft 26: Jagdpanzer IV, German Army and Waffen-SS Tank Destroyers, Western Front, 1944–1945.*
2. One battalion served in the defence of the Italian peninsula and a handful fought in the West as explained in the unit histories.
3. Although the vehicle was officially referred to as Hornisse until January 1944, I have used the term Nashorn throughout this section to avoid confusion.

A Nashorn self-propelled gun, possibly from 3.Kompanie, of schwere Panzerjäger-Abteilung 560 probably photographed in late 1943 or early 1944 when the battalion was fighting in southern Ukraine. The bracket for the spare roadwheel, situated on the superstructure oblique behind the jack, the Hummel-style gun support and the drive sprocket would suggest that this is one of the vehicles that were allocated to the battalion in early 1943. The large L-shaped fittings welded to the superstucture side towards the rear are non-standard and provision has also been made to hold lengths of track on the lower hull sides. Note the Winterketten tracks on the hull front.

From the earliest days of the Russian Campaign the Axis armies in the East were controlled by large Army Groups as shown here: Heersgruppe Nord (A); Heeresgruppe Mitte (B),and; Heeresgruppe Süd (C). By the spring of 1943 the Red Army had forced the Germans back from the banks of the Don River, west of Stalingrad, entering the city of Kursk on 8 February and Kharkov, the second-largest city in Ukraine, just over a week later. Although Kharkov was retaken in March, the front line had been pushed back in the south from the Caucasus Mountains to what is today central Ukraine as shown by the shaded area (1) between Orel and Tikorhetsk. On the front of Heeresgruppe Nord the siege of Leningrad, which had been encircled since September 1941, had effectively been lifted by the end of January 1943. The last German troops were evacuated from the Demyansk Pocket (2) on 28 February 1943 and the Rzhev Pocket was abandoned by the end of March (3). The dark broken line (4) indicates the front line on the eve of the German summer offensive, codenamed Operation Citadel, which commenced on 5 July 1943. The main German Armies are also shown in the positions they held at that time. Following the suspension of Operation Citadel the Russians launched several major counteroffensives which eventually forced the Germans to withdrawn to the western bank of the Dnieper River by the end of the year (5).

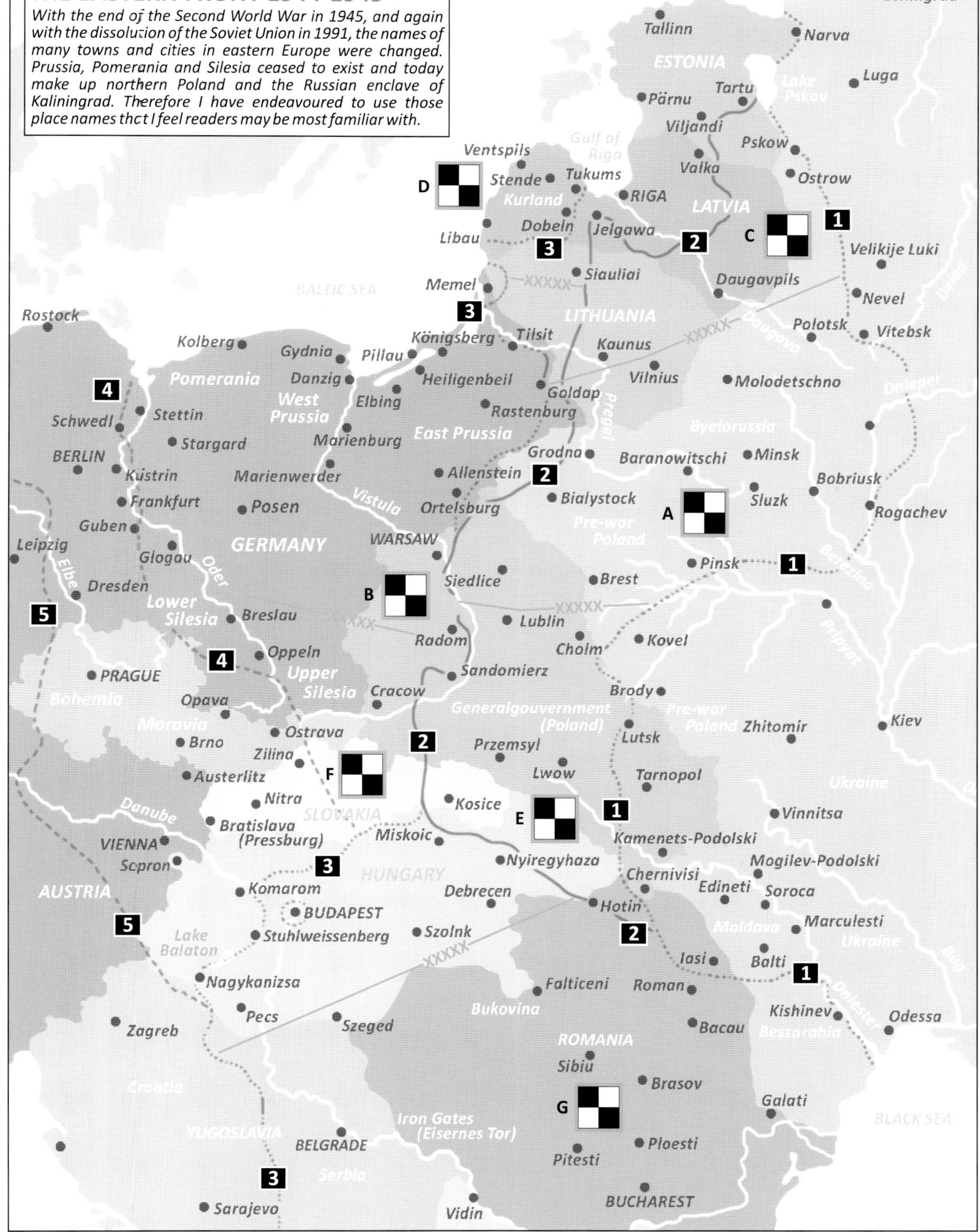

During the first months of 1944 the Soviets made significant gains on the northern sector but these early assaults would be dwarfed by their summer offensive, codenamed Operation Bagration, which commenced on 22 June. Heeresgruppe Mitte (A) was largely destroyed in the summer battles of 1944 and although it was rebuilt to some extent it could not hope to hold the extensive front line that had been created by the Soviet advance. The divisions of Heeresgruppe Mitte now operated between northern Poland and southern Lithuania (B). Heeresgruppe Nord (C) was now restricted to the Baltic States and was eventually cut off there (D). The southern half of the front, formerly Heeresgruppe Süd (E), was now controlled by Heeresgruppe Nordukraine (F) and Heeresgruppe Südukraine (G). The first broken line (1) indicates the front line as it was on the evening of 21 June 1944 just prior to the commencement of Operation Bagration. By mid-August 1944 the Red Army had advanced to the Vistula (2) and the second broken line (3) shows the front as it was on the last day of the year. The rapid expansion in the south was due in large part to the surrender of Romania and Bulgaria. 4. 30 March 1945. 5. 11 May 1945, the demarcation line agreed by the United States, Britain and the Soviet Union.

A Nashorn of Stab, schwere Panzerjäger-Abteilung 519 said to have been photographed in southern Belarus during the summer of 1944. Note the unit insignia on the superstructure oblique and the method of holding the spare roadwheel. The name Berlin is just visible on the superstructure side and the naming of the battalion's vehicles is discussed further on page 8. This Nashorn is fitted with the Pzkpfw III ausf H drive sprocket and has the external gun travel lock introduced in May 1943. Partly obscured by the canvas cover is the curved shield of the 8.8cm gun and metal reinforcing loop. Note the large stowage box on the right side front fender and the position of the towing cable.

Initially it was planned that the Nashorn self-propelled guns would be allocated to the Panzerjäger battalions of selected Panzer divisions. In company-sized units these powerful weapons would support the battalion's smaller calibre guns, acting as long-range tank killers. But this idea was soon discarded, perhaps at Hitler's instigation, and it was decided that they would operate as semi-independent battalions, each made up of three companies, controlled at army level.

Before examining the histories of these units I should explain some of the classifications used in wartime documents. The original Nashorn battalions, numbered 93, 519, 525, 560 and 655, were raised between March and August 1943 and in the official orders which have survived the war they are referred to as 'schw.Panz.Jäg.Abteilung', 's.Panz.Jäg.Abteilung (Hornisse)' and 'schw.Panz.Jäg.Abteilung (8,8cm Pz.Jag. IV b Hornisse)'. In one instance the first two titles are employed in the same document. In the Zuführungsliste, the vehicle allocation lists compiled between May 1943 and April 1945, all battalions are referred to simply as Pz.Jg.Abt. Similarly, the KstN authorised for the Nashorn battalions use the titles Pz.Jäg.Abt and Pz.Jäg.Kp.'Hornisse' (see also page 60). But from early 1944 the term schwere Heeres-Panzerjäger-Abteilung (sHPzJg-Abt) was increasingly used by the battalions, particularly those numbered 88, 93, and 525, the latter confirmed by a set of award documents dated April and June 1944. All these units were of course army-level troops and the term 'Heeres' would seem appropriate. But a number of battalions continued to use their old title, noticeably sPzJgAbt 519 and sPzJgAbt 560, the former recorded on the OKW situation maps for mid-1944 and the latter entered on a Wehrpass and dated 1 June 1944. In view of this seeming confusion, and as a matter of convenience, I have referred to all these battalions as schwere Panzerjäger-Abteilung, or sPzJg-Abt, but mentioned any name changes where they can be confirmed. Similarly I have referred to the self-propelled 8.8cm vehicle as Nashorn (pronounced Nahz-horn), or Rhinoceros, although it was officially known as Hornisse, or Hornet, until 27 January 1944 when Hitler directed that the name should be changed. Interestingly, the name Hornisse appears in many official reports and lists for some months after that date. I have also used the abbreviations sPzJg-Kp for schwere Panzerjäger-Kompanie, PzJg-Ers-Abt for Panzerjäger-Ersatz-und Ausbildungs-Abteilung, Insp.d PzTr for Inspekteur der Panzertruppen and HZA for Heereszeugamt, the office of the Heereswaffenamt (HWA) responsible for the allocation of armoured vehicles.

For the sake of completeness, and to help the reader make sense of the allocation details reproduced on pages 15 and 16, I have also included brief summaries regarding the units which served on the Western Front and in Italy.

sPzJg-Abt 88. In late September 1943 the companies of Pz.Jg.Abt 88 found themselves operating in a semi-independent role when the battalion's parent formation, 18.Panzer-Division (Pz-Div), was disbanded. From October the remnants of the battalion supported 18.Panzergrenadier-Division (Pzgr-Div) which was at that time fighting in the area to the south-east of Smolensk near the present-day Belarus border. But by late November 1943 the battalion was withdrawn from the front and sent to Mielau in Poland to be to rebuilt and was officially renamed sPzJg-Abt 88 on 20 December 1943 (1).

The allocation lists mention that the vehicles shipped to the battalion on Friday, 14 January 1944 completed its authorisation and during February and March the Nashorns of sPzJg-Abt 88 were transported to the Eastern Front. It was probably at about this time that the title sHPzJg-Abt came into use. Here the battalion was subordinated to 1.Panzerarmee and took part in the battles for Kamianets-Podilskyi in what is today western Ukraine. In May 1944 the battalion claimed to be the first German unit to destroy a Soviet IS-2 heavy tank but a number of accounts state that the vehicle had been captured by sPz-Abt 506, a Tiger unit, and was under tow at the time (2).

In July the battalion was attached to Korpsgruppe Balck, an ad-hoc battle group formed from parts of 8.Pz-Div, 18.SS-PzGren-Div and a number of infantry formations. It is probable that the companies operated independently at this time, attempting to blunt a Soviet breakthrough south of Lemberg. In late late August 1944 a Kampfgruppe scraped together from elements of 8.Pz.Div with ten Nashorns of sPzJg-Abt 88 was operating in the area between Radomysl Wielki and Tarnow, east of Krakow, and these may have been the battalion's only combat-ready Nashorns.

By January 1945 the battalion was fighting near Preiswitz in Silesia, modern-day Przyszowice, with the tanks of Pz-Regt 21 and some accounts state that at this time a number of Nashorns were used to test infrared night vision equipment. I have been unable to confirm this and it is highly unlikely that any were used in combat. In March the battalion was subordinated to 17.Armee which was operating near Wroclaw in western Poland and during that month received the last Nashorns allocated by the HZA.

By April 1945 the battalion had retreated into the area north-east of Prague where the surviving crews surrendered to the Soviets.

SPzJg-Abt 93. On 16 June 1943 the companies of PzJg-Abt 93, the anti-tank element of 26.Pz-Div which was at that time training in France, received orders to convert to the Nashorn self-propelled anti-tank gun. There is some indication that the battalion was to remain with its parent formation but the division was rushed to the Italian Front and on 23 July the battalion was formally detached and completed its training under the direction

.........text continued on page 7

Notes

1. Truppenubüngsplatz Mielua, situated at modern-day Mlawa, was one of three large training centres that were created in occupied Poland. Specialising in armoured and combined arms operations, it was jokingly referred to by locals as Nowy Berlin, or New Berlin.

2. I have been unable to confirm this curious story but there is a mention of an IS-1, a similar but smaller tank, in Oberst Wolfgang Schneider's exhaustive account of the Tiger battalions. This vehicle was indeed captured by sPz-Abt 506 on 1 May 1944 while the battalion was operating in the area north-west of Chernivtsi in support of 17.Pz-Div as was sPzJg-Abt 88.

Said to have been photographed near Orel, south-east of Bryansk, during the summer of 1943, these vehicles are from 1.Kompanie, schwere Panzerjäger-Abteilung 88, identified by the black club symbol on the superstructure side. Note also the number 06 painted on the closest vehicle on the lower edge of the superstructure towards the rear. Just visible are the coaxial machine guns that were fitted to most of the battalion's self-propelled guns. Examples are shown in the Camouflage & Markings section. By the summer of 1944 three-digit company numbers, rendered in a dark colour with a white outline, were in use with 1.Kompanie and 3.Kompanie at least but they are quite obviously absent here.

With the onset of winter, all the Nashorns of schwere Panzerjäger-Abteilung 88 were camouflaged with a coat of whitewash. The coverage varied from comprehensive to patchy and the headquarters vehicles were painted in random wavy lines or broad bands of white. In most cases the markings were avoided.

..........text continued from page 5

of 21.Pz-Div. An OKH order of 28 August 1943 refers to the battalion, perhaps for the first time, as sPzJg-Abt 93. By the first week of September the battalion was fully equipped with forty-five Nashorns and shortly after that was sent to the Eastern Front and subordinated to 6.Armee.

Here sPzJg-Abt 93 helped to cover the general withdrawal behind the Dniepr River and subsequently took part in the attack to retake Krivoi Rog, present-day Kryvyi Rih in central Ukraine. In early 1944 the battalion supported the tanks of 24.Pz-Div in the fighting between Nikolaiev and Jassy in eastern Romania and remained with the units of 6.Armee throughout the summer. In August the battalion was pushed back as far as Kichinew, today Chisinau in Moldova, and the crossings of the Pruth River and was almost completely destroyed in the battles there.

Records show that the battalion was reforming at Potsdam in November 1944 with PzJg-Ers-Abt 3 and it was anticipated that it would be rebuilt with Jagdpanther tank destroyers and Sturmgeschütz III assault guns.

But this proved to be impossible and by 1 December 1944 just twelve Nashorns, which had been hurriedly scraped together, were delivered to the crews of 1.Kompanie. These men found themselves seconded to a Kampfgruppe commanded by Major Hannibal von Lüttichau with 1.Kompanie, sPzJg-Abt 525 and elements of 7.Kompanie, Panzer-Regiment 2 (see Panzer-Regiment 2 below).

sPzJg-Abt 519. Formed on 25 August 1943 at Spremberg under the direction of PzJg-Ers-Abt 43 and originally built around Arbeitstab Hoppe (1), the battalion contained a headquarters, a headquarters company and three Nashorn companies.

In September 1943 the battalion was transferred to the military training area at Oldebroek in the Netherlands and the first Nashorn tank destroyers were received in the following month. The allocation lists for November 1943 mention that the battalion was fully equipped and preparations were made to move to the Eastern Front.

Subordinated to 3.Panzerarmee, which was at that time fighting to hold a bridgehead on the eastern bank of the Daugava River at Vitebsk in north-eastern Belarus, elements of the battalion directly supported 21.Pz-Div. The 8.8cm guns proved their worth here and in a single action fought on Christmas Eve the Nashorns of 1.Kompanie managed to disable thirty Soviet tanks. The commander of the company's first platoon, Leutnant Albert Ernst (2) and his crew, accounted for fourteen enemy vehicles supposedly expending just twenty-one rounds.

The battalion remained in this area until the early months of 1944, claiming that by the end of February a total of 290 Russian tanks had been knocked out for a loss of six Nashorns, four of which were destroyed by their own crews. During the rapid Soviet advances made in late June

..........text continued on page 10

Notes

1. Much has been made of this unit in some accounts but it was merely a temporary headquarters commanded by Major Wolf-Horst Hoppe who later led the battalion.
2. Later promoted to Hauptmann, Ernst led one of the last armoured units to surrender to the Western Allies and is mentioned in *TankCraft 42: Jagdtiger, Heavy Tank Destroyer, German Army, Western Front 1945.*

Oberleutnant, later Hauptmann, Alfred Ziemann, the commander of 1.Kompanie, schwere Panzerjäger-Abteilung 93, posing with one of the men of his company. Ziemann was awarded the Ritterkreuz, visible here, in January 1944 suggesting that this photograph was taken in the spring of that year when the battalion was fighting in Romania.

Both photographs on this page show Nashorns of schwere Panzerjäger-Abteilung 519 in action during the fighting for Vitebsk in late 1943. The battalion's Nashorns carried individual names on the superstructure sides and on the rear access door as can be seen here. The exact significance of the names is unclear but it is likely that Pommern, Brandenburg, depicted above, and Berlin were the three vehicles assigned to the battalion headquarters. Other confirmed names are Büffel and Falke, of 1.Kompanie, and Tiger and Puma, probably from 3.Kompanie. A possible contender for 2.Kompanie is Bismarck. These names are examined further in the Camouflage & Markings section. Note also the battalion's unit insignia on the superstructure rear plate.

Probably photographed at about the same time as the vehicles depicted on the facing page, this Nashorn of schwere Panzerjäger-Abteilung 519 is fitted with the purpose-built gun travel lock and support, a single Bosch headlight and brackets for a spare roadwheels on either side of the hull front. Although the paint is badly worn, the battalion's unit insignia is just visible on the right side, as viewed, of the superstructure oblique. Below that is the tactical symbol identifying a self-propelled anti-tank company.

Said to have been photographed in Belarus during the summer of 1944, this Nashorn named Berlin is the same vehicle depicted on page 4. It seems likely that when the winter whitewash camouflage was removed in the spring of 1944 many of the battalion's vehicles were left in a scheme of plain Dunkelgelb without any disruptive pattern. The power of the 8.8cm gun is shown to good effect in this image as the muzzle blast removes a large portion of the hut's roof.

..........text continued from page 7

the battalion was separated from 3.Panzerarmee and supported the units of 4.Armee which was fighting in the area north of Orscha. By 1 July 1944 sPzJg-Abt 519 had been reduced to as few as fifteen operational Nashorns and these vehicles, together with a number of Sturmgeschütz III assault guns, were formed into Panzergruppe Hoppe.

By August the surviving crews were withdrawn to Mielau in Poland where the battalion was rebuilt with Jagdpanther tank destroyers and Sturmgeschütz III assault guns (1). The remaining vehicles were handed over to s.Pz.Jg.Abt 664 (see below).

sPzJg-Abt 525. Formed in June 1943, the battalion was sent to the Italian Front in the following August with a full complement of forty-five Nashorns. The battalion took part in the defence of Cassino and the counterattacks at Anzio where the crews claimed that they were able to destroy at least one Sherman tank at a range of 2,800 metres. From early 1944 the battalion was referred to as sHPzJg-Abt 525.

On 9 September 1944 the battalion's 1.Kompanie was detached and sent to Mielau to begin its conversion to Jagdpanther tank destroyers but this never took place (see Panzer-Regiment 2).

The first company's remaining vehicles were distributed between 2.Kompanie and 3.Kompanie which soldiered on in Italy where, with a handful of Nashorns on strength, they surrendered to units of the US Army in April 1945. The plan to equip 1.Kompanie with Jagdpanthers was eventually abandoned and the crews were assigned ten Nashorns in November 1944 and spent the last months of the war fighting on the Western Front (see Panzer-Regiment 2 below).

SPzJg-Abt 560. On 22 March 1943 OKH ordered that 3.Kompanie, PzJg-Abt 37, 4.Kompanie, PzJg-Abt 41 and 4.Kompanie, PzJg-Abt 42, which were at that time undergoing training in France, should be used to form sHPzJg-Abt 560 and outfitted with the Nashorn self-propelled anti-tank gun.

By the second week of May the battalion was fully equipped but training was extremely slow due to a lack of basic supplies which meant that gunnery practice, and even driver instruction, had to be curtailed until the battalion arrived in Russia. As late as 17 June 1943 the battalion was not considered combat ready and it did not take a direct part in Operation Citadel as had originally been planned. Subordinated to Armee Abteilung Kempf (2) the battalion secured the right flank of III.Panzerkorps and was subsequently engaged in the fierce defensive battles of late 1943.

At least one authoritative source states that in November and December 1943 the battalion was assigned a number of Marder III self-propelled guns but this is not borne out by the surviving allocation lists. By the end of the year the battalion claimed to have destroyed 251 Soviet tanks. In the first weeks of 1944 the battalion took part in the attempt to recapture Kirovograd, modern-day Kropyvnytskyi, but was so badly depleted by April that it was withdrawn from the front and sent to Mielau to be re-equipped with Jagdpanther tank destroyers (3).

Notes

1. There is some debate regarding the assault guns and this may have been a mixed unit which also contained some tanks.
2. Confusingly referred to as an Abteilung, possibly due to Hitler's personal dislike of its commander General Werner Kempf, this army-sized battle group was made up of three corps and various support units. Referred to in some accounts as Panzergruppe Kempf.
3. The battalion's history is also examined in *TankCraft 8: Jagdpanther Tank Destroyer, German Army and Waffen-SS, Western Europe 1944-1945.*

Fgst Nr.30030, assembled in March 1943, was the first Nashorn to fall into enemy hands. Assigned to 1.Kompanie, schwere Panzerjäger-Abteilung 560, it is depicted here after its capture by the Soviets. Note the early drive sprocket, the armoured boxes which cover part of the the exhaust pipe and behind them the large muffler. Of note is the additional armour plate bolted to the gun shield which was an identifying feature of this battalion. These features are examined in detail in the Technical Details and Modifications section.

A Nashorn of 1.Kompanie, schwere Panzerjäger-Abteilung 560 probably photographed in late summer 1943 when the battalion had been in combat for some time as evidenced by the kill rings on the barrel of the 8.8cm gun. This vehicle was assembled before May 1943 and is fitted with the gun travel lock and support originally designed for the Hummel self-propelled howitzer. Note the spare tracks on the hull side.

sPzJg-Abt 655. Formed in early April 1943 with a new headquarters company and the remnants of PzJag-Kp 521, PzJag-Kp 611 and PzJag-Kp 670, this battalion was the second unit to receive supplies of the Nashorn self-propelled anti-tank gun (1).

The first elements of the battalion were transported to the East in May and began live-fire exercises in the Mogilev area. Problems with the Zielgerät 34 gun sight were experienced from the first day when it was found that the support for the sight was too weak to compensate for the recoil of the 8.8cm gun and even cross-country driving necessitated that the sight needed to be constantly adjusted, which was impossible under combat conditions. Just thirteen vehicles of the full complement were fitted with the improved Zielgerät 1a which rectified the problem. To add to the battalion's woes a Soviet air attack made during the evening of Monday, 21 June 1943 completely destroyed five Nashorns and severely damaged a further five. In addition, nine crewmen were killed and eighteen wounded. On the following Friday PzJag-Kp 611 and PzJag-Kp 670 were sent to Minsk and ordered to await replacements for the lost vehicles and new gun sights but both companies were stranded there until the end of July.

The thirteen vehicles equipped with the Zielgerät 1a gun sight and the crews of PzJag-Kp 521 were subordinated to 2.Armee and for most of July 1943 supported XXXV.Armeekorps in attempting to halt the Russian breakout from the Kursk salient. Fighting between Orel and the Dnieper River, the company claimed the destruction of almost sixty Soviet tanks and a number of transport vehicles. But two Nashorns were written off while twenty-eight men were wounded with another two killed, serious casualties for a company-sized unit. On 27 July 1943 the headquarters element and the three companies were united in Minsk. In November the battalion was again sent to the front and was attached to LVI.Panzerkorps which was operating in the Pripyat Marsh area in what is today northern Ukraine near the Belarus border (2).

Throughout the early weeks of 1944 the battalion remained with LVI.Panzerkorps operating as a combat reserve with the Tigers of sPz-Abt 505 and a number of Sturmgeschütz units.

On 2 April 1944 the companies were at last renamed as 1., 2. and 3.Kompanie, sPzJag-Abt 655. In August the battalion, less 3.Kompanie, was withdraw from the front and sent to Mielau in central Poland to begin training on the Jagdpanther tank destroyer and later served on the Western Front until the end of the war.

The battalion's 3.Kompanie, with the remaining twenty-four Nashorns, was detached and subordinated directly to 4.Panzerarmee. This company, initially referred to as Einsatz-Kompanie 655, served as an operational reserve in the Sandomierz area until late November 1944 (see sPzJg-Komp 669 below).

Notes

1. The companies were referred to by these numbers for some time after the battalion's formation. In some accounts the original unit is referred to as sPzJg-Abt Stalingrad but the official order for its creation makes no mention of it.
2. Variously known as the Pinsk Marsh, the Pripet Marsh, the Polesie Marsh and the Rokitno Marsh. This huge natural wetland is fed by the Pripyat River.

sHPzJg-Abt 664. Originally formed in late September 1943 from elements of PzJg-Abt 263 and PzJg-Abt 291, the battalion's three companies were all equipped with 8.8cm Pak 43 towed anti-tank guns. The battalion served on the Eastern Front, subordinated to 3.Panzerarmee, but lost most, if not all, the guns during June and July 1944.

On 18 August a total of twelve Nashorns were handed over from sPzJg-Abt 519 which was returning to Germany (1) and in early September it was proposed that the battalion be reorganised with 1.Kompanie and 3.Kompanie equipped with towed anti-tank guns while the Nashorns would be concentrated in 2.Kompanie. In January 1945 a report authored by Pionier-Regimentsstab z.b.V. 623, a headquarters unit attached to 4.Armee, shows that a number of Nashorns were still on hand with the battalion but their combat effectiveness was regarded as limited due to what was described as their 'Empfindlichkeit', or sensitivity. Note that all the battalions raised to operate the towed 8.8cm 43 anti-tank gun were officially referred to as schwere Heeres-Panzerjäger-Abteilungen from May 1943 when the first unit was formed.

sPzJg-Komp 93. See Pz-Rgt 2.

sPzJg-Komp 669. Formed on 25 November 1944 from 3.Kompanie, sPzJg-Abt 655 (see above). By January of the following year the company was operating with 17.Pz-Div and during the early winter battles was almost completely destroyed. Rebuilt in early February 1945 with thirteen new vehicles, the company remained with 17.Pz-Div until the end of the war, taking part in battles in northern Czechoslovakia.

sPzAbt 424. This battalion was equipped with Tiger II tanks (2) but lost all its heavy equipment in the fighting between Lisow and Kielce and the withdrawal to Czestochowa in little more than a week in mid-January 1945. The surviving crews, for the most part from 1.Kompanie and

A Nashorn of schwere Panzerjäger-Abteilung 655 photographed during the summer of 1943. This is an early production vehicle which lacks the reinforcing loop for the gun shield and is fitted with the initial gun support and lock. Note the battalion's unit insignia on the left side fender, as viewed, and the tactical symbol denoting a self-propelled anti-tank unit on the right. What may be a small number 1 next to the latter, which would indicate the company, is just visible.

Notes

1. Interestingly, both units had earlier been the subject of an extensive comparison undertaken by the headquarters of 3.Panzerarmee to assess the benefits of the towed gun against those of the self-propelled version.
2. Referred to as sPzAbt 501 until 21 December 1944 when the name was changed. The Tiger units that served in the East in 1945 are examined in *Tank Craft 31: Tiger I and Tiger II Tanks, German Army and Waffen-SS, The Last Battles in the East 1945.*

2.Kompanie, were later assembled at Grünberg, present-day Zielona Gora in western Poland, where they were used to form a company-sized Kampfgruppe equipped with three Pzkpfw IV and two Panther tanks, two Nashorn tank destroyers and a number of Jagdpanzer 38 Hetzers which were taken over from the nearby repair facilities at Brieg. This Kampfgruppe took part in the fighting on the Oder Front, principally around Reichthal and Linden, but was officially deactivated on 11 February 1945 and its personnel used to form sPzJg-Abt 512 (1).

1.Pz-Div. Strength reports for 6.Armee dated 30 December 1944 indicate that this division had a single operational Nashorn on hand with a further three in repair. There is no record in the official allocation lists of any vehicles being assigned to this formation but the report does seem to be accurate. At least two Kriegsgliederungen, or organisational charts, complied by the division show that 2.Kompanie, PzJg-Abt 37 was equipped with four Nashorns in November 1944 which were still on hand in January 1945 (2).

In the following month the battalion was transferred to Neuhammer in Germany for retraining and all four vehicles were apparently taken over by the regimental headquarters of the division's Panzer-Artillerie-Regiment 73 and assigned to that unit's I.Abteilung in March 1945.

2.SS-Pz-Div. A report compiled for Heeresgruppe B states that this division had a total of twenty Nashorns on hand, with twelve of those in repair, on 30 December 1944. There is no mention in the official allocation lists of any vehicles being shipped to this formation and the only Nashorns operating on the Western Front in late 1944 were the vehicles of 1.Kompanie, sPzJg-Abt 93 and 1.Kompanie, sPzJg-Abt 525 (see Pz-Regt 2) (3).

Panzer-Regiment 2. A rather enigmatic entry in the official allocation list for November 1944 gives a total of twenty-two Nashorns assigned to 'II.2' (see page 15). In the same month the headquarters of II.Abteilung, Panzer-Regiment 2 was detached from its parent formation, which was at that time fighting in the East, and transferred to Niefern in Germany. Here it was joined by 1.Kompanie, sPzJg-Abt 525 and 1.Kompanie, sPzJg-Abt 93 and on Wednesday, 29 November 1944 ten Nashorns were shipped to the former while another twelve vehicles were sent to the latter on the following Friday.

Officially at least these vehicles were assigned to II.Abteilung, Panzer-Regiment 2, hence 'II.2'. With the arrival of the regiment's 7.Kompanie these units were formed into Jagdpanzer-Abteilung von Lüttichau (4) which was involved in the fighting on the Western Front under the tactical command of Panzer-Brigade 106. On 27 December the two companies were detached and subordinated to sPzJg-Abt 654, which was operating in the Herrlisheim area, and later supported 198.Infanterie-division and Sturmgeschütz-Brigade 280 during Operation Sonnenwende (5).

In late February 1945 the surviving Nashorns were amalgamated into a single company which may have been referred to as sPzJg-Komp 93. The company could field just ten vehicles but one of these managed to disable an American M26 Pershing tank outside the village of Glesch, west of Cologne, with a single shot fired at close range (6). Most of the crews surrendered in the Ruhr Pocket and although a small number managed to escape and went on to form Pz-Div Clausewitz, it is certain that all the vehicles were lost. The personnel of 2.Kompanie may also have been used to build this division but records from this stage of the war are understandably confused.

Panzer-Kompanie Kummersdorf. This ad-hoc formation was created on 31 March 1945 and was to be built around the assets of Versuchs-Abteilung Waffenamt A, a unit based at the weapons trial centre at Kummersdorf south-west of Berlin. The company's three Panzer platoons were a very mixed bag and were reported to contain a Jagdtiger, a Tiger II, four Panthers, several obsolete and captured tanks and one Nashorn (7). On 19 April the company was moving south towards Luckau, about 60 kilometres from Berlin, but two days later it was ordered to join Kampfgruppe Möws which at that time was at Zossen, just outside the capital. Although it is impossible to be certain, this vehicle may be one of the Nashorns shipped to the Heereswaffenamt for testing in June 1943.

PzJg-Ers-Ausb-Abt 3. Located at Potsdam, just outside Berlin, this training and replacement unit was responsible for sPzJg-Abt 93 and sPzJg-Abt 560. In May 1943 a single Nashorn was assigned here from the Insp.d PzTr.

PzJg-Ers-Ausb-Abt 43. Established at Spremberg in early 1940, this unit provided training and replacements for sPzJg-Abt 519 and sPzJg-Abt 655. In May 1943 two Nashorns were assigned here from the Insp.d PzTr. This unit and PzJg-Ers-Ausb-Abt 3 were activated by the mobilisation of training units in January 1945 and sent to the area around Glogau on the Oder Front but is not known if they took any of the Nashorns that had been allocated with them.

Panzertruppen-Schule Wünsdorf. This establishment, which provided tactical and technical training for the army's tank troops, was renamed several times and eventually moved from Wünsdorf to Bergen. In May 1943 two Nashorns were assigned here from the Insp.d PzTr but their ultimate fate is unknown.

Notes

1. The history of this battalion is examined in *TankCraft 42: Jagdtiger, Heavy Tank Destroyer, German Army, Western Front 1945.*

2. It will be remembered that a company of this battalion had been used to raise sPzJg-Abt 560.

3. There may have been some plan to subordinated the two Nashorn companies to the Waffen-SS division or allocate new vehicles. The suggestion that the entry was a simple clerical error seems unlikely as all the others from the 30 December report have been confirmed. There is certainly no mention of these twenty Nashorns when the division took part in the fighting in northern Hungary in February 1945.

4. Major von Lüttichau's unit was later joined by a Hetzer company.

5. A favourite codename employed during the war and therefore somewhat confusing. This assault was a follow-up to Operation Nordwind.

6. Among the many controversies surrounding this event is the contention, mainly in American accounts, that the German unit in question was 2.Kompanie. sPzJg-Abt 93.

7. This unit is examined in some detail in *TankCraft 31:Tiger I and Tiger II Tanks, German Army and Waffen-SS, The Last Battles in the East 1945.*

HORNISSE/NASHORN ALLOCATIONS, 1943-1945

The allocation of armoured vehicles was the responsibility of the various Heereszeugämter (singular, Heereszeugamt or HZA), or army ordnance offices, located throughout Germany, Austria and also at Danzig and Posen, present-day Gdansk and Poznan in Poland. The HZA reported to the Heereswaffenamt (HWA), the army's ordnance directory which oversaw all aspects of weapons development and supply, and vehicles designated as Panzer or Panzerjäger were assigned by the HZA at the request of the Generalinspekteur der Panzertruppen. Sometimes, but rarely, vehicles were dispatched directly from the manufacturer as an expedient. Information relating to the allocation and transport of each shipment was recorded in Führungsliste, or supply lists, and provision was made on these lists to record the unit to which vehicles were

Month and Year	Unit	Quantity	Shipped	Date	Remarks
May 1943	sPzJg-Abt 655	10	6	7.5	Authorised allocation of 45 vehicles completed with this shipment (1)
			4	14.5	
	sPzJg-Abt 560	15	6	14.5	Authorised allocation of 45 vehicles completed with this shipment (2)
			9	17.5	
	Insp.d.PzTr	5	2	20.5	PzJg-Ers-Abt 43
			1	18.5	PzJg-Ers-Abt 3
			2	21.5	Panzertruppen-Schule Wünsdorf
	sPzJg-Abt 525	5	3	22.5	
			2	2.6	
June 1943	Heereswaffenamt	1	1	15.6	
	sPzJg-Abt 525	40	5	2.6	
			10	19.6	
			10	1.7	
			8	2.7	
	PzJg-Ers-Abt 43	1	1	22.6	
July 1943	sPzJg-Abt 525	7	7	10.7	Mention of 5 vehicles shipped in May and a further 33 in June (3)
	sPzJg-Abt 655	8	8	27.7	
	sPzJg-Abt 93	31	16	29.7	
	sPzJg-Abt 93		10	4.8	
August 1943	Insp.d.PzTr	9	1	9.8	
			2	10.8	
			1	13.8	
			3	23.8	
			2	31.8	
	sPzJg-Abt 93	19	10	18.8	Authorised allocation of 45 vehicles completed with this shipment
			9	10.9	
September 1943	Nachschub Ost (4)	10	5	30.9	To sPzJg-Abt 560
October 1943	sPzJg-Abt 560	10	10	31.10	Received 14.10 (5)
	Nachschub Ost	10	5	31.10	Received 14.10 by sPzJg-Abt 655 (5)
			5	31.10	Received 14.10 by sPzJg-Abt 560 (5)
	sPzJg-Abt 519	20	6	31.10	
			13	6.11	Received 14.10 (5)
	sPzJg-Abt 655	5	5	13.11	
	Wa Prüf 8 (6)	1	1	8.11	
	sPzJg-Abt 519	5	5	6.11	

1. An allocation of thirty-five vehicles must have been made at some time between the first week of April 1943, when the battalion was formed, and the end of the month. 2. By this time the battalion's 1.Kompanie and 3.Kompanie had received thirty Nashorns, presumably in April, and had arrived in Russia. Therefore we can assume that the May allocation was for 2.Kompanie. 3. The June reference here confirms the shortfall recorded for that month. 4. Nachschub Ost was the main ordnance replacement authority for the Eastern Front. 5. I have shown the dates received here as they are entered on the official list although they cannot be correct and most sources suggest that these vehicles actually arrived at their destination in November. 6. Wa Prüf 8, or Waffen Prüfwesen 8, was a section of the HWA responsible for testing of, among other things, optics.

assigned, the number of vehicles, the date the allocation was confirmed, the date transport was arranged and the date on which the shipment was actually dispatched. There were also columns for railway shipping numbers and to advise the date that the shipment arrived at its destination but in many instances these were, unfortunately, left blank. Nevertheless, the information that is included is remarkably accurate, and the chart below was compiled largely from the Hornisse/Nashorn Führungsliste. I am unable to explain why the lists begin in May 1943 when at least sixty-five vehicles had been delivered by the end of the previous month and as many as eighty-five examples had left the assembly lines. Note that this chart includes all allocations for the period in question including those made to units serving on the Western and Southern Fronts.

Month and Year	Unit	Quantity	Shipped	Date	Remarks
November 1943	sPzJg-Abt 519	11	5		Authorised allocation of 45 vehicles completed with this shipment
			6		
	sPzJg-Abt 560	5	5		
	sPzJg-Abt 93	5	5		
	sPzJg-Abt 655	5	5		
	PzJg-Abt	12	9	9.12	Notation '1.Res PzJg-Abt 5' deleted by hand (7)
			2	9.12	
December 1943	sPzJg-Abt 655	34	4	9.12	
			11	19.12	
			3	6.1	
			15	14.1	Authorised allocation of 45 vehicles completed with this shipment
			1	14.1	
January 1944	sPzJg-Abt 93	10	10	30.1	
	sPzJg-Abt 560	5	4	3.2	
February 1944	sPzJg-Abt 655	10	10	10.3	
March 1944	sPzJg-Abt 655	10	5	22.3	
			5	1.4	
April 1944	sPzJg-Abt 93	10	5	3.5	
May 1944	sPzJg-Abt 525	5	5	26.5	
	sPzJg-Abt 88	10	10	17.6	
	sPzJg-Abt 93	10	10	5.7	
	sPzJg-Abt 519	5	5	12.6	
June 1944	sPzJg-Abt 93	10	10	1.8	25 vehicles on hand (vorhanden) (8)
July 1944	No allocations recorded for this month (9)				
August 1944	sPzJg-Abt 88	30	10	9.8	
			10	1.9	Received 5.9
			10	26.8	Received 5.9
September 1944	sPzJg-Abt 525	20	15	13.9	Received 18.9
			5	23.9	
October 1944	sPzJg-Abt 525 (10)	20	10	15.10	Received 29.10
			10	9.11	Received 14.11
November 1944	II.Abt/PzRegt 2	22	10	29.11	
December 1944	No allocations recorded for this month (11)				
January 1945	sPzJg-Kp 669 (12)	4	4	13.1	
February 1945	sPzJg-Abt 669 (13)	13	13	10.2	Received 10.2
March 1945	sPzJg-Abt 88	4	4	11.3	Received 13.3

7. Reserve-Panzerjäger-Abteilung 5 had been created in August 1943 and subordinated to 155.Reserve-Panzer-Division which was later absorbed by 9.Panzer-Division. 8. This figure would seem to be at odds with the allocations made previously but it is quite clear in the list and has in fact been changed by hand from twenty-seven. 9. Each month of the Führungsliste is given a separate page and all pages are numbered consecutively which confirms that none are missing. 10. The name Nashorn was used in the allocation lists for the first time in this month. 11. A handwritten notation 'December keine Verteilung', or no allocations for December, made on the November 1944 list is also typed on the December page. 12. For some reason this allocation is included on the December 1944 page as a separate entry. 13. Although referred to as a battalion in the February list, this unit is in fact schwere Panzerjäger-Kompanie 669.

A Nashorn of 1.Kompanie, schwere Panzerjäger-Abteilung 519 photographed at some time after 7 February 1944 with Leutnant Albert Ernst, one of the company's platoon commanders, standing in the superstructure on the right. On 23 December 1943, in the fighting around Vitebsk, Ernst and his crew were credited with destroying fourteen Soviet tanks and holding back a major enemy assault. Leutnant Ernst was awarded the Ritterkreuz on 22 January 1944 but the actual ceremony did not take place until the following month, hence the date given for our photograph. Several technical points to note are the wooden jack block on the right side trackguard, the method of storing the towing cable, the inner surface of the driver's visor, the enlarged air intakes and single headlight. Note also that the upper section of the gun travel lock remains in place when the supporting struts are lowered.

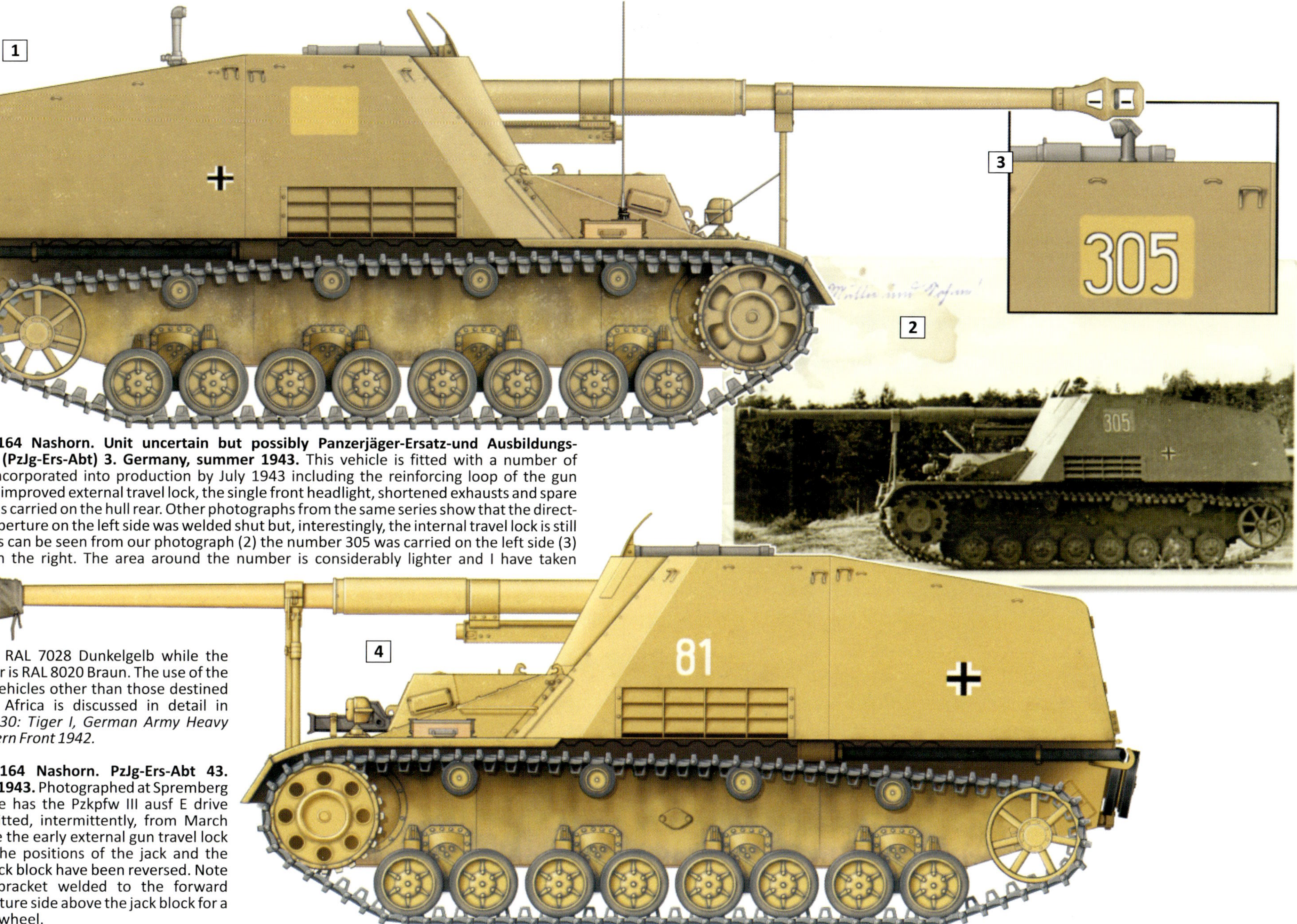

1. Sdkfz 164 Nashorn. Unit uncertain but possibly Panzerjäger-Ersatz-und Ausbildungs-Abteilung (PzJg-Ers-Abt) 3. Germany, summer 1943. This vehicle is fitted with a number of features incorporated into production by July 1943 including the reinforcing loop of the gun shield, the improved external travel lock, the single front headlight, shortened exhausts and spare roadwheels carried on the hull rear. Other photographs from the same series show that the direct-fire sight aperture on the left side was welded shut but, interestingly, the internal travel lock is still in place. As can be seen from our photograph (2) the number 305 was carried on the left side (3) but not on the right. The area around the number is considerably lighter and I have taken this to be RAL 7028 Dunkelgelb while the base colour is RAL 8020 Braun. The use of the latter on vehicles other than those destined for North Africa is discussed in detail in *TankCraft 30: Tiger I, German Army Heavy Tank, Eastern Front 1942.*

4. Sdkfz 164 Nashorn. PzJg-Ers-Abt 43. Germany, 1943. Photographed at Spremberg this vehicle has the Pzkpfw III ausf E drive sprocket fitted, intermittently, from March 1943. Note the early external gun travel lock and that the positions of the jack and the wooden jack block have been reversed. Note also the bracket welded to the forward superstructure side above the jack block for a spare roadwheel.

1

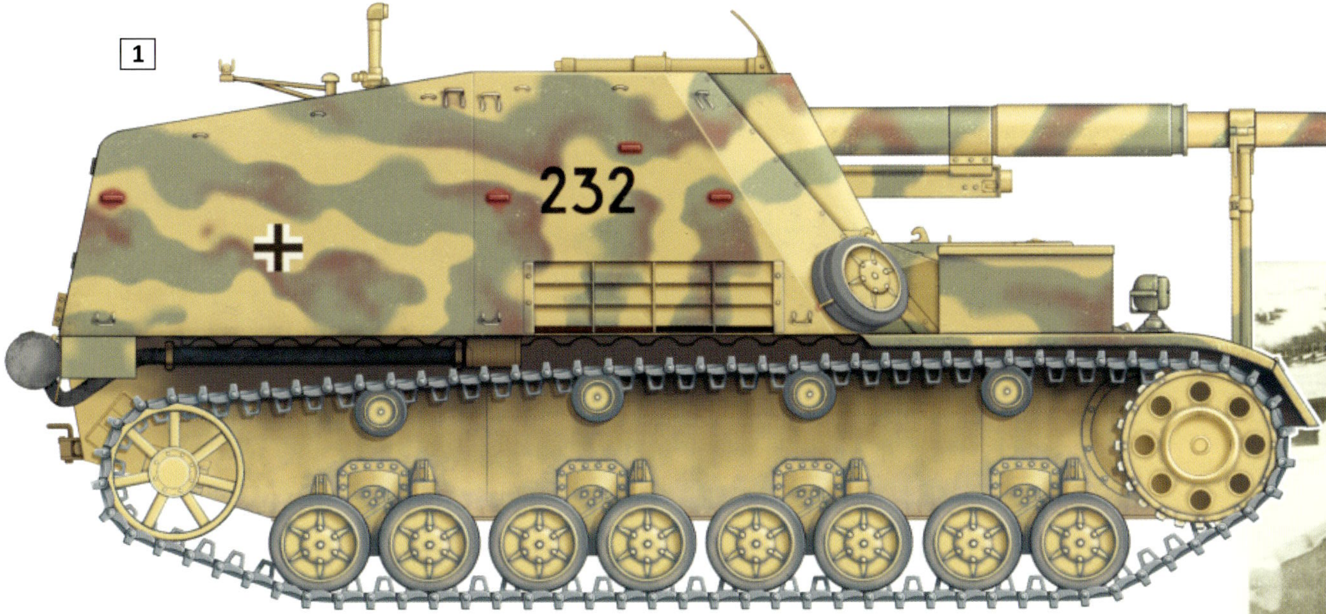

3

1. Sdkfz 164 Nashorn. 2.Kompanie, schwere Panzerjäger-Abteilung 560. Germany, April 1943. Photographed as the battalion was leaving for the Eastern Front, this vehicle is finished in a camouflage pattern typical of 2.Kompanie. Note that the broad bands of Olivgrün painted over the base of Dunkelgelb are interspersed with patches of Rotbraun. The placement and style of company number was common throughout the battalion, as can be seen in the illustrations on the following pages, and the large stowage box fitted to the right side trackguard is also seen on most of the battalion's vehicles. Note the non-standard brackets welded to the superstructure sides, here painted in Rot primer colour, the early exhaust with its armoured box and the Pzkpfw III ausf E drive sprocket.

2

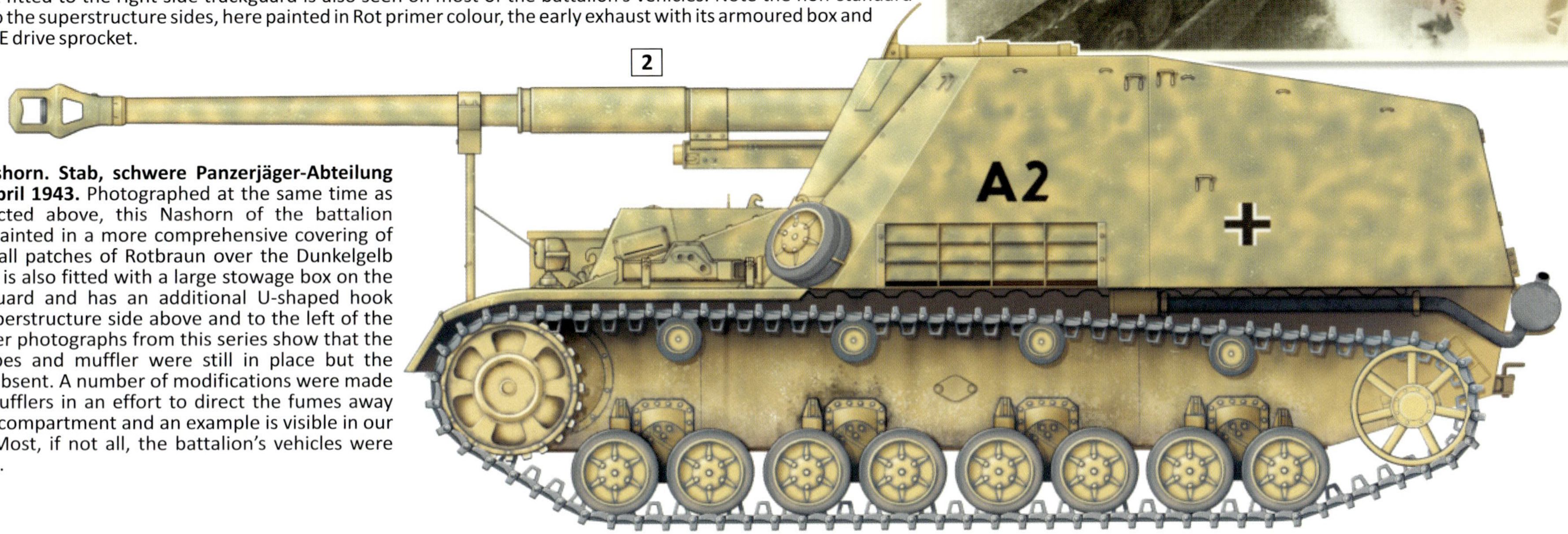

2. Sdkfz 164 Nashorn. Stab, schwere Panzerjäger-Abteilung 560. Germany, April 1943. Photographed at the same time as the vehicle depicted above, this Nashorn of the battalion headquarters is painted in a more comprehensive covering of Olivgrün with small patches of Rotbraun over the Dunkelgelb base. This vehicle is also fitted with a large stowage box on the right side trackguard and has an additional U-shaped hook welded to the superstructure side above and to the left of the Balkenkreuz. Other photographs from this series show that the early exhaust pipes and muffler were still in place but the armoured box is absent. A number of modifications were made to the exhaust mufflers in an effort to direct the fumes away from the fighting compartment and an example is visible in our photograph (3). Most, if not all, the battalion's vehicles were similarly modified.

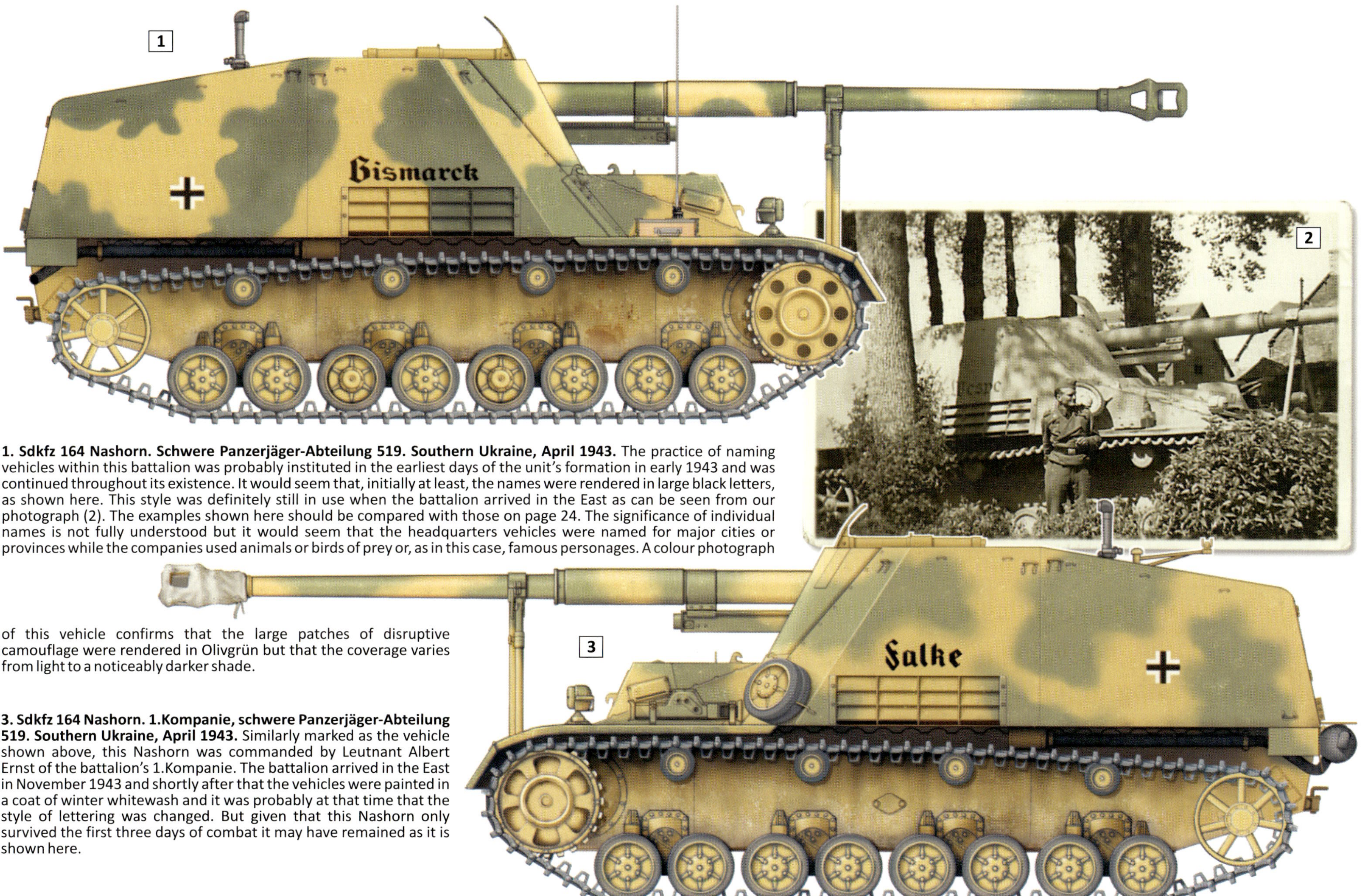

1. Sdkfz 164 Nashorn. Schwere Panzerjäger-Abteilung 519. Southern Ukraine, April 1943. The practice of naming vehicles within this battalion was probably instituted in the earliest days of the unit's formation in early 1943 and was continued throughout its existence. It would seem that, initially at least, the names were rendered in large black letters, as shown here. This style was definitely still in use when the battalion arrived in the East as can be seen from our photograph (2). The examples shown here should be compared with those on page 24. The significance of individual names is not fully understood but it would seem that the headquarters vehicles were named for major cities or provinces while the companies used animals or birds of prey or, as in this case, famous personages. A colour photograph of this vehicle confirms that the large patches of disruptive camouflage were rendered in Olivgrün but that the coverage varies from light to a noticeably darker shade.

3. Sdkfz 164 Nashorn. 1.Kompanie, schwere Panzerjäger-Abteilung 519. Southern Ukraine, April 1943. Similarly marked as the vehicle shown above, this Nashorn was commanded by Leutnant Albert Ernst of the battalion's 1.Kompanie. The battalion arrived in the East in November 1943 and shortly after that the vehicles were painted in a coat of winter whitewash and it was probably at that time that the style of lettering was changed. But given that this Nashorn only survived the first three days of combat it may have remained as it is shown here.

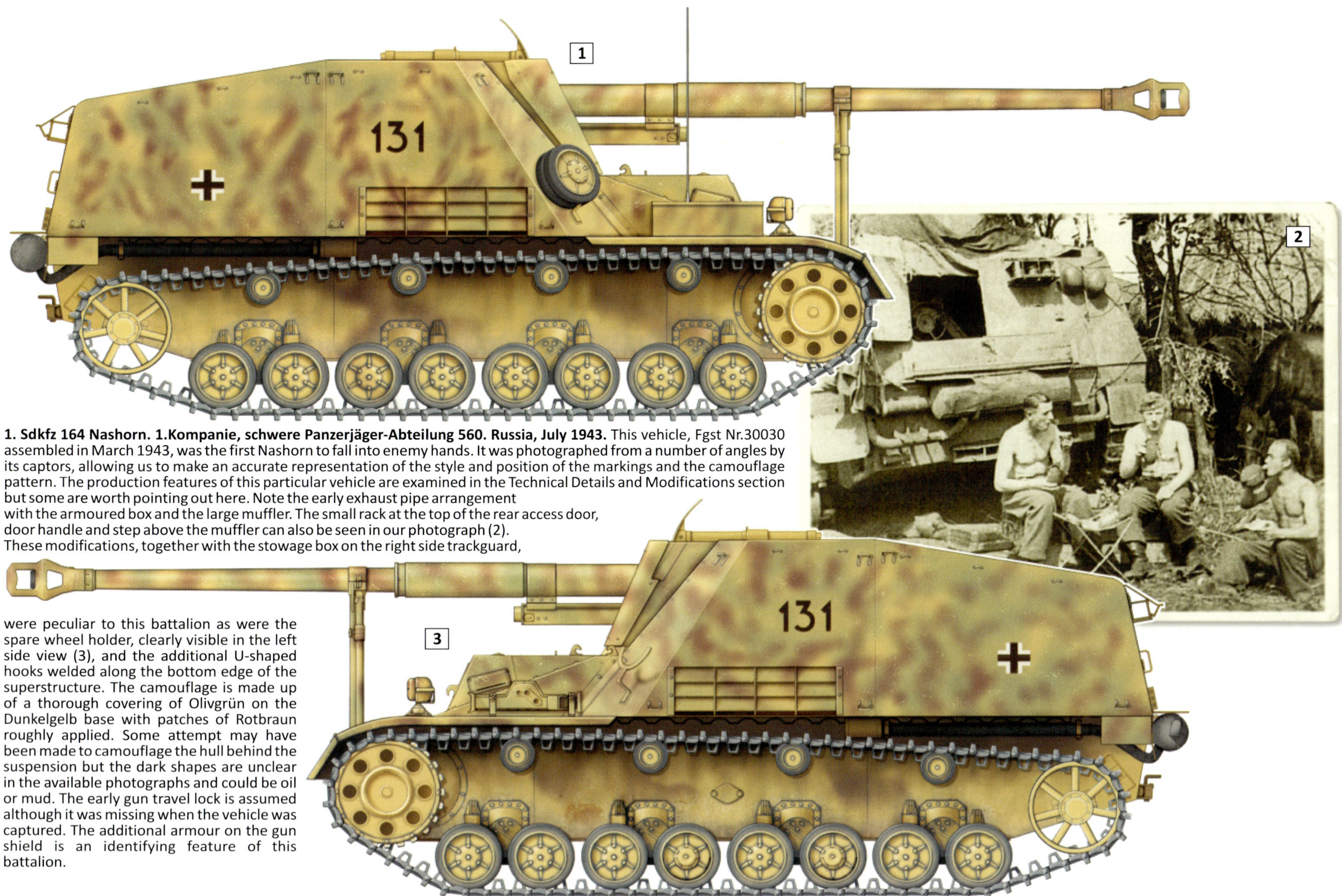

1. Sdkfz 164 Nashorn. 1.Kompanie, schwere Panzerjäger-Abteilung 560. Russia, July 1943. This vehicle, Fgst Nr.30030 assembled in March 1943, was the first Nashorn to fall into enemy hands. It was photographed from a number of angles by its captors, allowing us to make an accurate representation of the style and position of the markings and the camouflage pattern. The production features of this particular vehicle are examined in the Technical Details and Modifications section but some are worth pointing out here. Note the early exhaust pipe arrangement with the armoured box and the large muffler. The small rack at the top of the rear access door, door handle and step above the muffler can also be seen in our photograph (2). These modifications, together with the stowage box on the right side trackguard, were peculiar to this battalion as were the spare wheel holder, clearly visible in the left side view (3), and the additional U-shaped hooks welded along the bottom edge of the superstructure. The camouflage is made up of a thorough covering of Olivgrün on the Dunkelgelb base with patches of Rotbraun roughly applied. Some attempt may have been made to camouflage the hull behind the suspension but the dark shapes are unclear in the available photographs and could be oil or mud. The early gun travel lock is assumed although it was missing when the vehicle was captured. The additional armour on the gun shield is an identifying feature of this battalion.

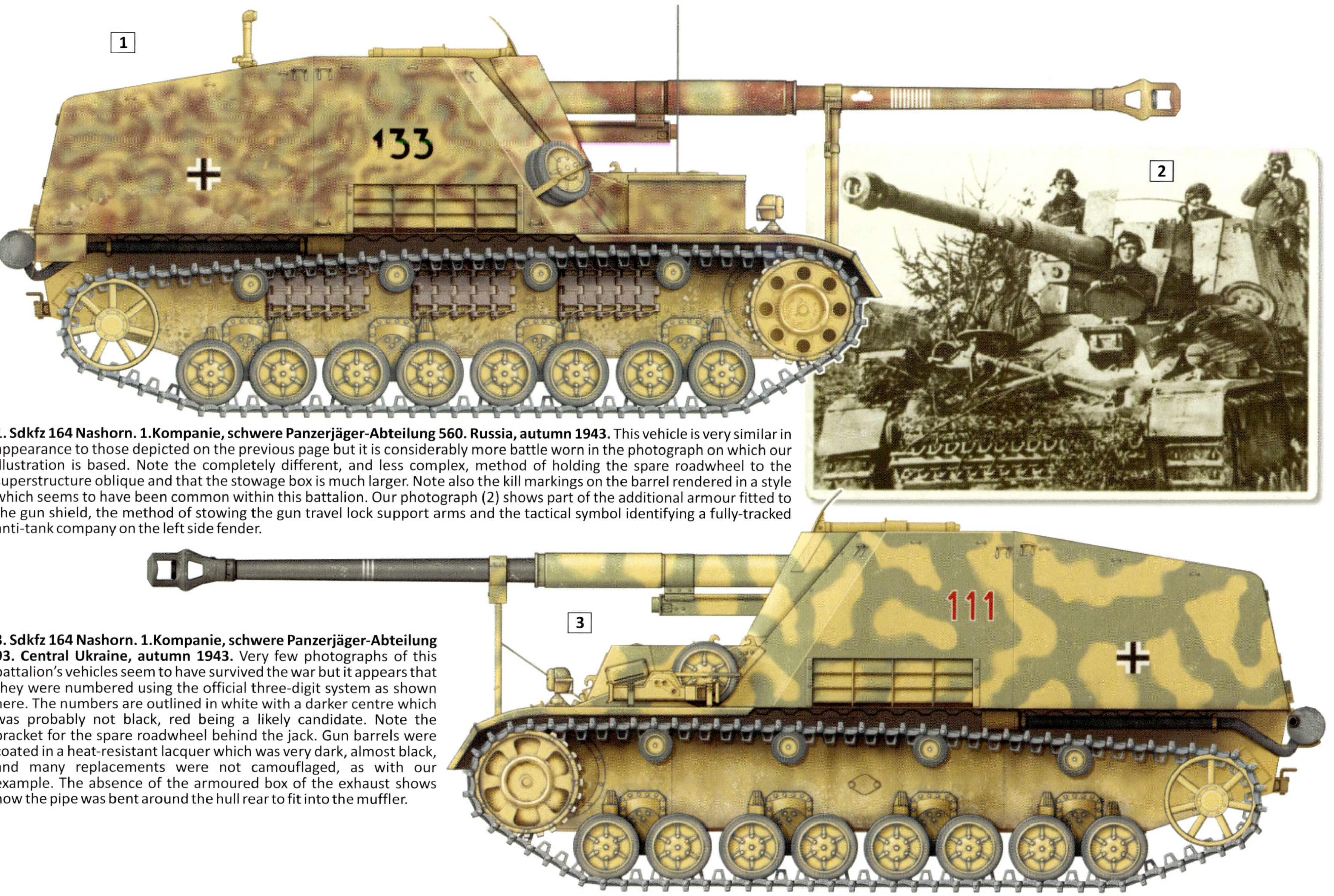

1. Sdkfz 164 Nashorn. 1.Kompanie, schwere Panzerjäger-Abteilung 560. Russia, autumn 1943. This vehicle is very similar in appearance to those depicted on the previous page but it is considerably more battle worn in the photograph on which our illustration is based. Note the completely different, and less complex, method of holding the spare roadwheel to the superstructure oblique and that the stowage box is much larger. Note also the kill markings on the barrel rendered in a style which seems to have been common within this battalion. Our photograph (2) shows part of the additional armour fitted to the gun shield, the method of stowing the gun travel lock support arms and the tactical symbol identifying a fully-tracked anti-tank company on the left side fender.

3. Sdkfz 164 Nashorn. 1.Kompanie, schwere Panzerjäger-Abteilung 93. Central Ukraine, autumn 1943. Very few photographs of this battalion's vehicles seem to have survived the war but it appears that they were numbered using the official three-digit system as shown here. The numbers are outlined in white with a darker centre which was probably not black, red being a likely candidate. Note the bracket for the spare roadwheel behind the jack. Gun barrels were coated in a heat-resistant lacquer which was very dark, almost black, and many replacements were not camouflaged, as with our example. The absence of the armoured box of the exhaust shows how the pipe was bent around the hull rear to fit into the muffler.

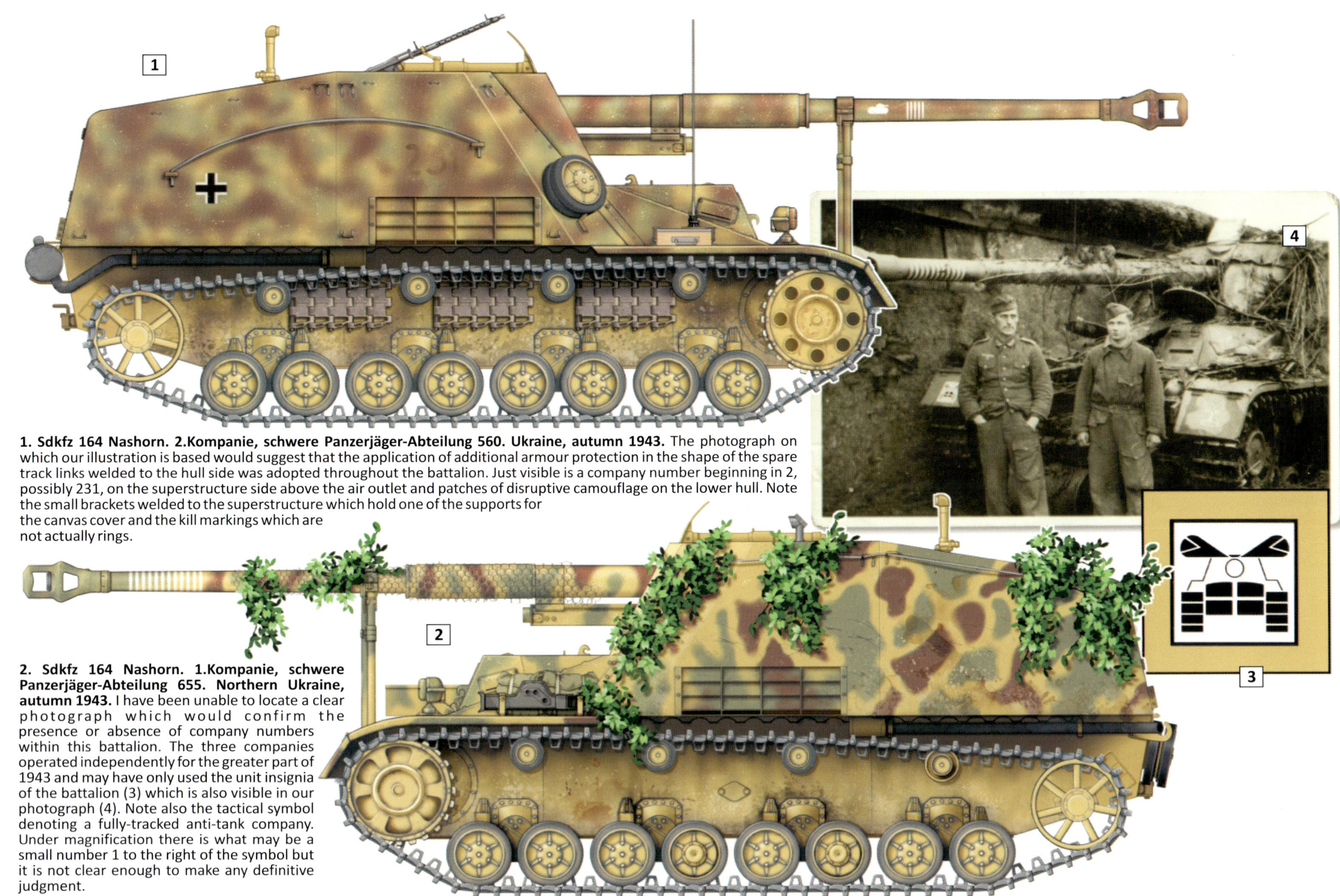

1. Sdkfz 164 Nashorn. 2.Kompanie, schwere Panzerjäger-Abteilung 560. Ukraine, autumn 1943. The photograph on which our illustration is based would suggest that the application of additional armour protection in the shape of the spare track links welded to the hull side was adopted throughout the battalion. Just visible is a company number beginning in 2, possibly 231, on the superstructure side above the air outlet and patches of disruptive camouflage on the lower hull. Note the small brackets welded to the superstructure which hold one of the supports for the canvas cover and the kill markings which are not actually rings.

2. Sdkfz 164 Nashorn. 1.Kompanie, schwere Panzerjäger-Abteilung 655. Northern Ukraine, autumn 1943. I have been unable to locate a clear photograph which would confirm the presence or absence of company numbers within this battalion. The three companies operated independently for the greater part of 1943 and may have only used the unit insignia of the battalion (3) which is also visible in our photograph (4). Note also the tactical symbol denoting a fully-tracked anti-tank company. Under magnification there is what may be a small number 1 to the right of the symbol but it is not clear enough to make any definitive judgment.

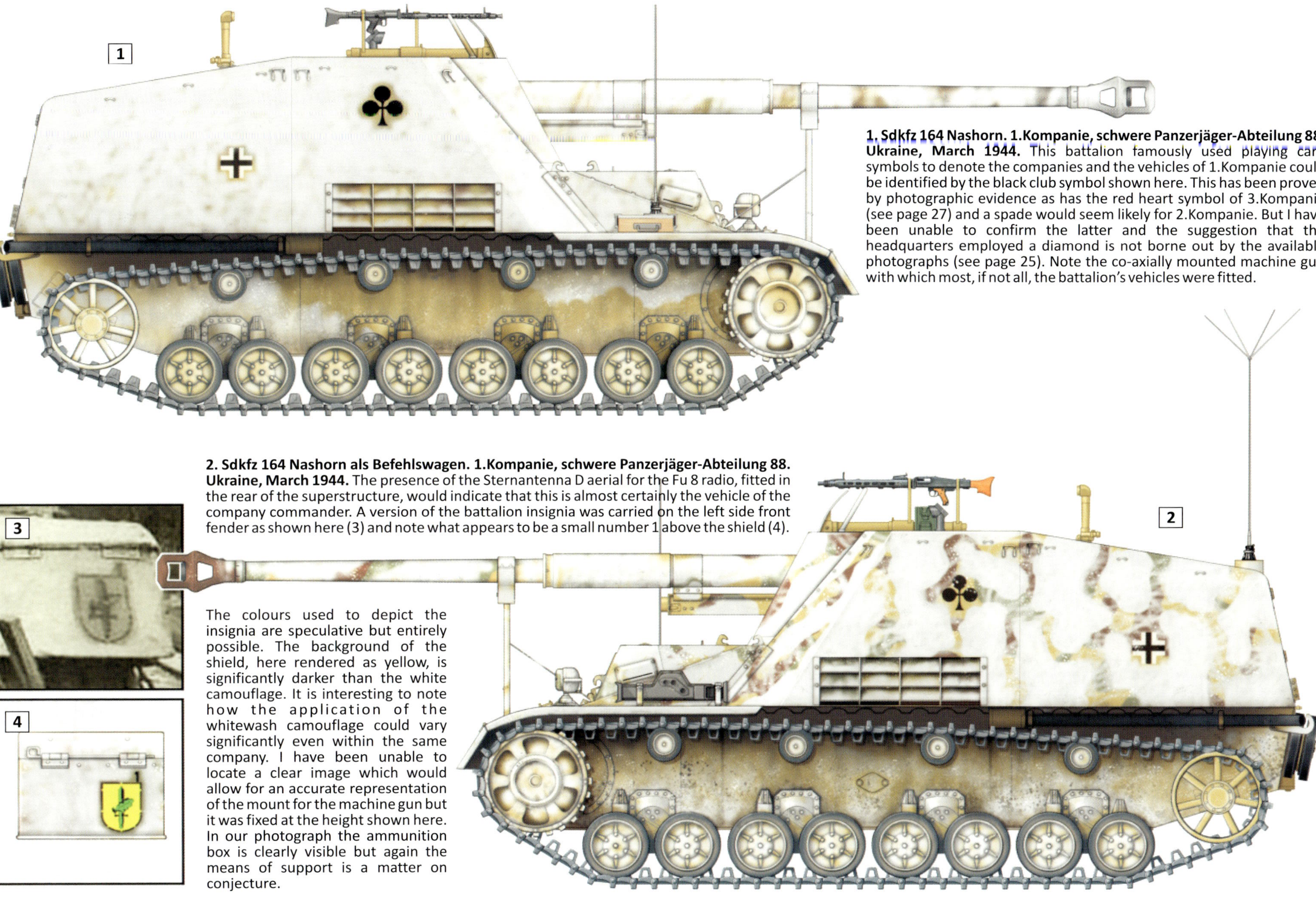

1. Sdkfz 164 Nashorn. 1.Kompanie, schwere Panzerjäger-Abteilung 88. Ukraine, March 1944. This battalion famously used playing card symbols to denote the companies and the vehicles of 1.Kompanie could be identified by the black club symbol shown here. This has been proven by photographic evidence as has the red heart symbol of 3.Kompanie (see page 27) and a spade would seem likely for 2.Kompanie. But I have been unable to confirm the latter and the suggestion that the headquarters employed a diamond is not borne out by the available photographs (see page 25). Note the co-axially mounted machine gun with which most, if not all, the battalion's vehicles were fitted.

2. Sdkfz 164 Nashorn als Befehlswagen. 1.Kompanie, schwere Panzerjäger-Abteilung 88. Ukraine, March 1944. The presence of the Sternantenna D aerial for the Fu 8 radio, fitted in the rear of the superstructure, would indicate that this is almost certainly the vehicle of the company commander. A version of the battalion insignia was carried on the left side front fender as shown here (3) and note what appears to be a small number 1 above the shield (4).

The colours used to depict the insignia are speculative but entirely possible. The background of the shield, here rendered as yellow, is significantly darker than the white camouflage. It is interesting to note how the application of the whitewash camouflage could vary significantly even within the same company. I have been unable to locate a clear image which would allow for an accurate representation of the mount for the machine gun but it was fixed at the height shown here. In our photograph the ammunition box is clearly visible but again the means of support is a matter on conjecture.

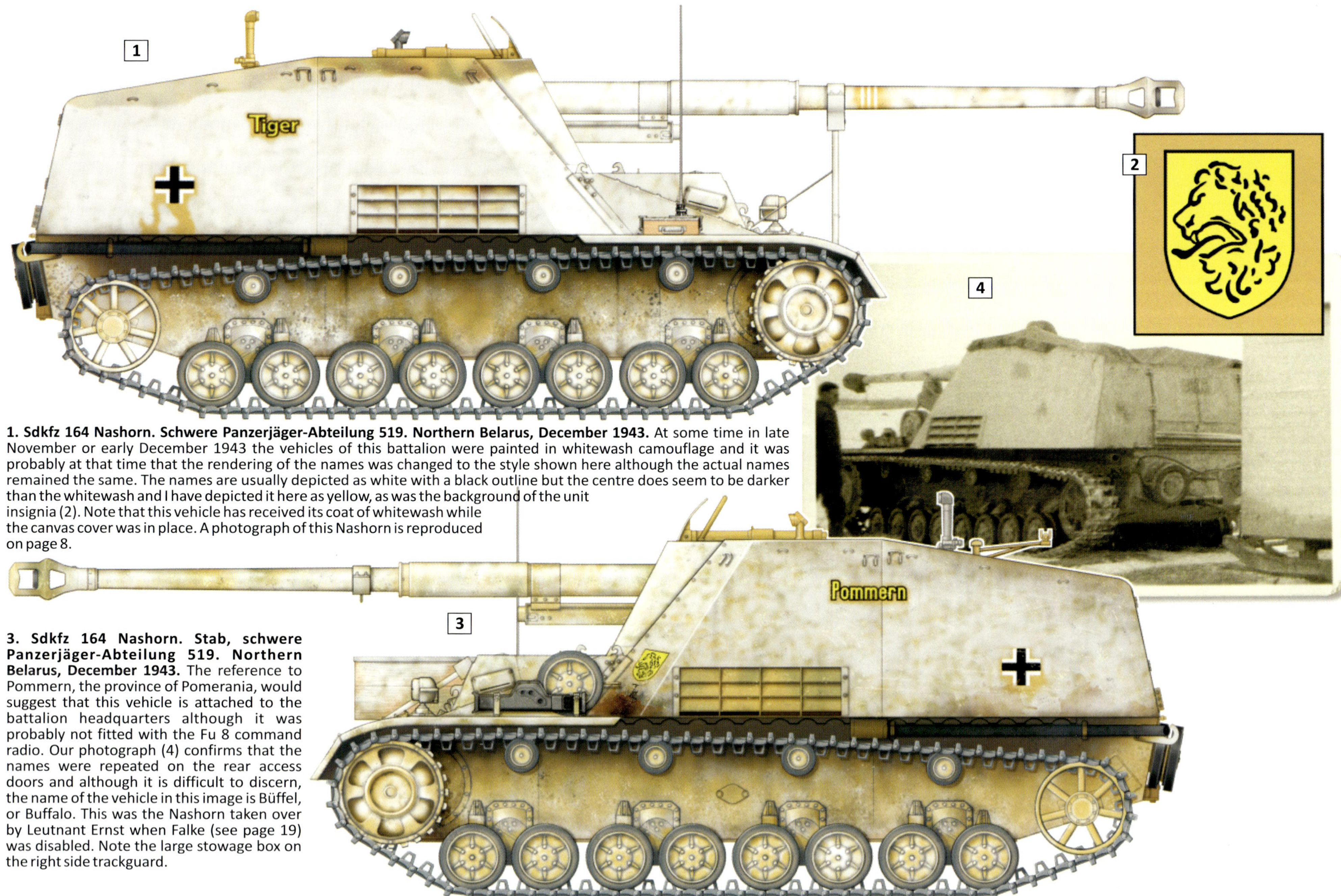

1. Sdkfz 164 Nashorn. Schwere Panzerjäger-Abteilung 519. Northern Belarus, December 1943. At some time in late November or early December 1943 the vehicles of this battalion were painted in whitewash camouflage and it was probably at that time that the rendering of the names was changed to the style shown here although the actual names remained the same. The names are usually depicted as white with a black outline but the centre does seem to be darker than the whitewash and I have depicted it here as yellow, as was the background of the unit insignia (2). Note that this vehicle has received its coat of whitewash while the canvas cover was in place. A photograph of this Nashorn is reproduced on page 8.

3. Sdkfz 164 Nashorn. Stab, schwere Panzerjäger-Abteilung 519. Northern Belarus, December 1943. The reference to Pommern, the province of Pomerania, would suggest that this vehicle is attached to the battalion headquarters although it was probably not fitted with the Fu 8 command radio. Our photograph (4) confirms that the names were repeated on the rear access doors and although it is difficult to discern, the name of the vehicle in this image is Büffel, or Buffalo. This was the Nashorn taken over by Leutnant Ernst when Falke (see page 19) was disabled. Note the large stowage box on the right side trackguard.

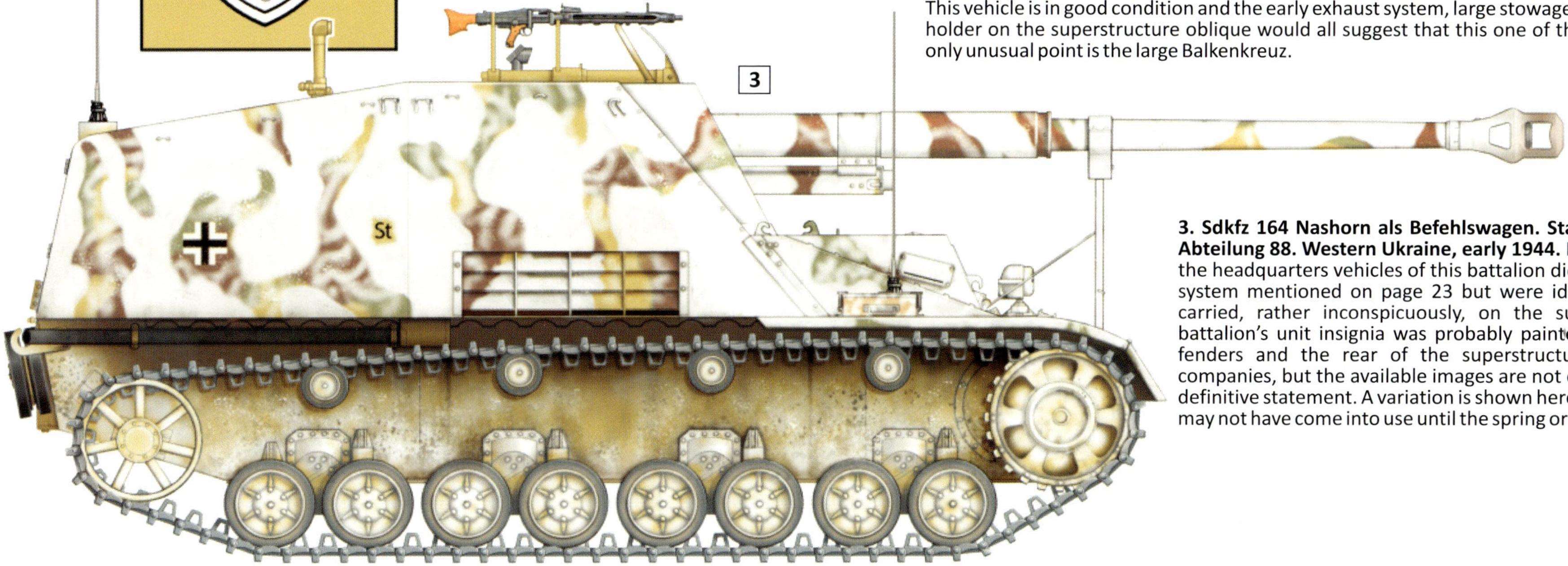

1. Sdkfz 164 Nashorn. 1.Kompanie, schwere Panzerjäger-Abteilung 560. Ukraine, late 1943 or early 1944. By late November or early December 1943 the vehicles of this battalion had been camouflaged with whitewash and most were comprehensively covered as can be seen from our photograph (2). This vehicle is in good condition and the early exhaust system, large stowage box and spare roadwheel holder on the superstructure oblique would all suggest that this one of the original allocation. The only unusual point is the large Balkenkreuz.

3. Sdkfz 164 Nashorn als Befehlswagen. Stab, schwere Panzerjäger-Abteilung 88. Western Ukraine, early 1944. Photographs indicate that the headquarters vehicles of this battalion did not use the playing card system mentioned on page 23 but were identified by the letters St carried, rather inconspicuously, on the superstructure sides. The battalion's unit insignia was probably painted onto one of the front fenders and the rear of the superstructure, as it was with the companies, but the available images are not clear enough to make any definitive statement. A variation is shown here (4) but I suspect that this may not have come into use until the spring or summer of 1944.

1. Sdkfz 164 Nashorn. Stab, schwere Panzerjäger-Abteilung 519. Southern Belarus, spring 1944. When the winter camouflage was removed many, but not all, the vehicles of this battalion appeared as our illustration in a plain base coat of Dunkelgelb. The black outline of the name was also not universal and another example, Brandenburg, can be seen on page 8. Note that the unit insignia is of a slightly smaller size (2) than that shown on page 24. I have depicted the background of the shield, and the name Berlin, as white as it is decidedly lighter than the Dunkelgelb base in our photograph but I admit that is largely conjecture on my part.

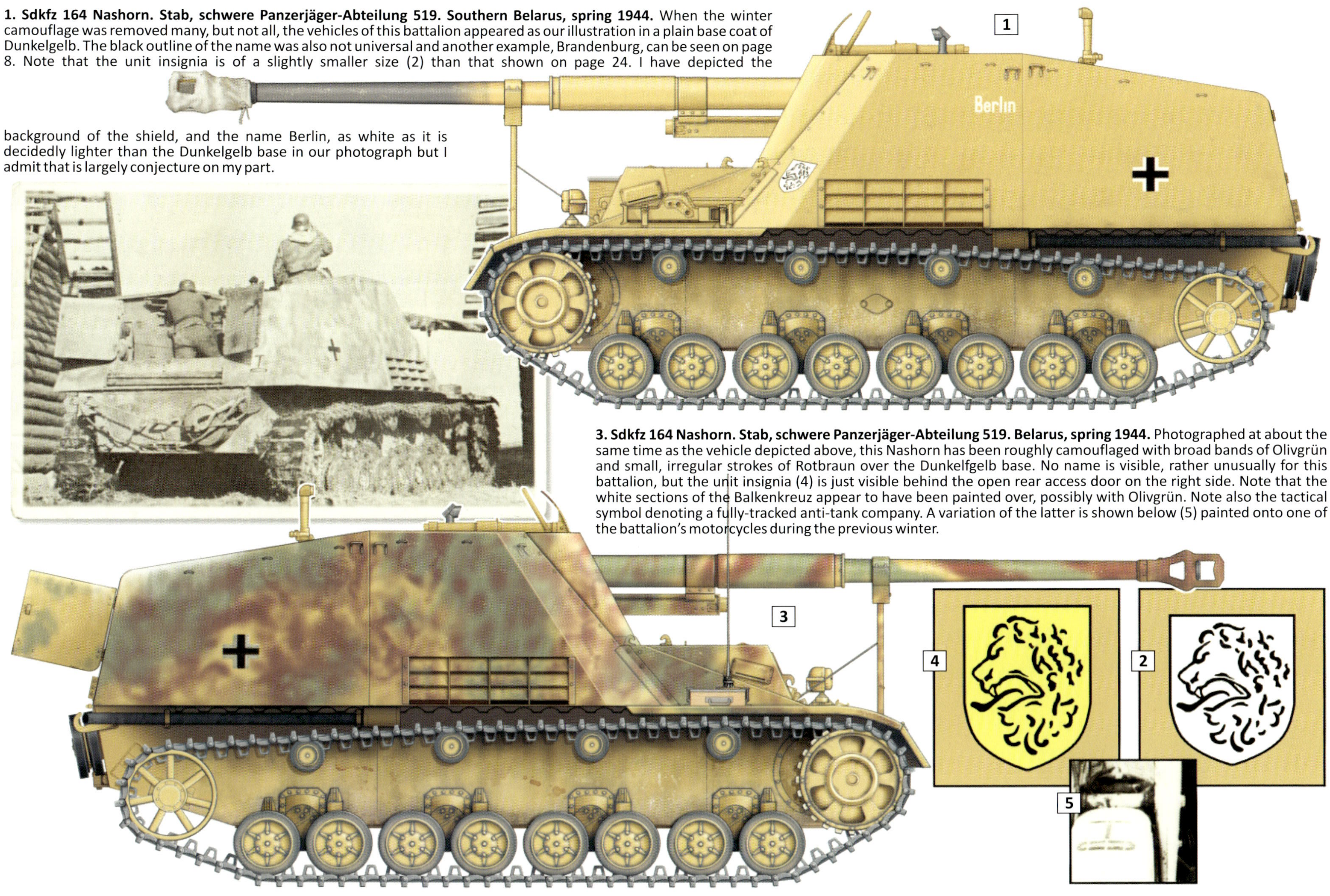

3. Sdkfz 164 Nashorn. Stab, schwere Panzerjäger-Abteilung 519. Belarus, spring 1944. Photographed at about the same time as the vehicle depicted above, this Nashorn has been roughly camouflaged with broad bands of Olivgrün and small, irregular strokes of Rotbraun over the Dunkelfgelb base. No name is visible, rather unusually for this battalion, but the unit insignia (4) is just visible behind the open rear access door on the right side. Note that the white sections of the Balkenkreuz appear to have been painted over, possibly with Olivgrün. Note also the tactical symbol denoting a fully-tracked anti-tank company. A variation of the latter is shown below (5) painted onto one of the battalion's motorcycles during the previous winter.

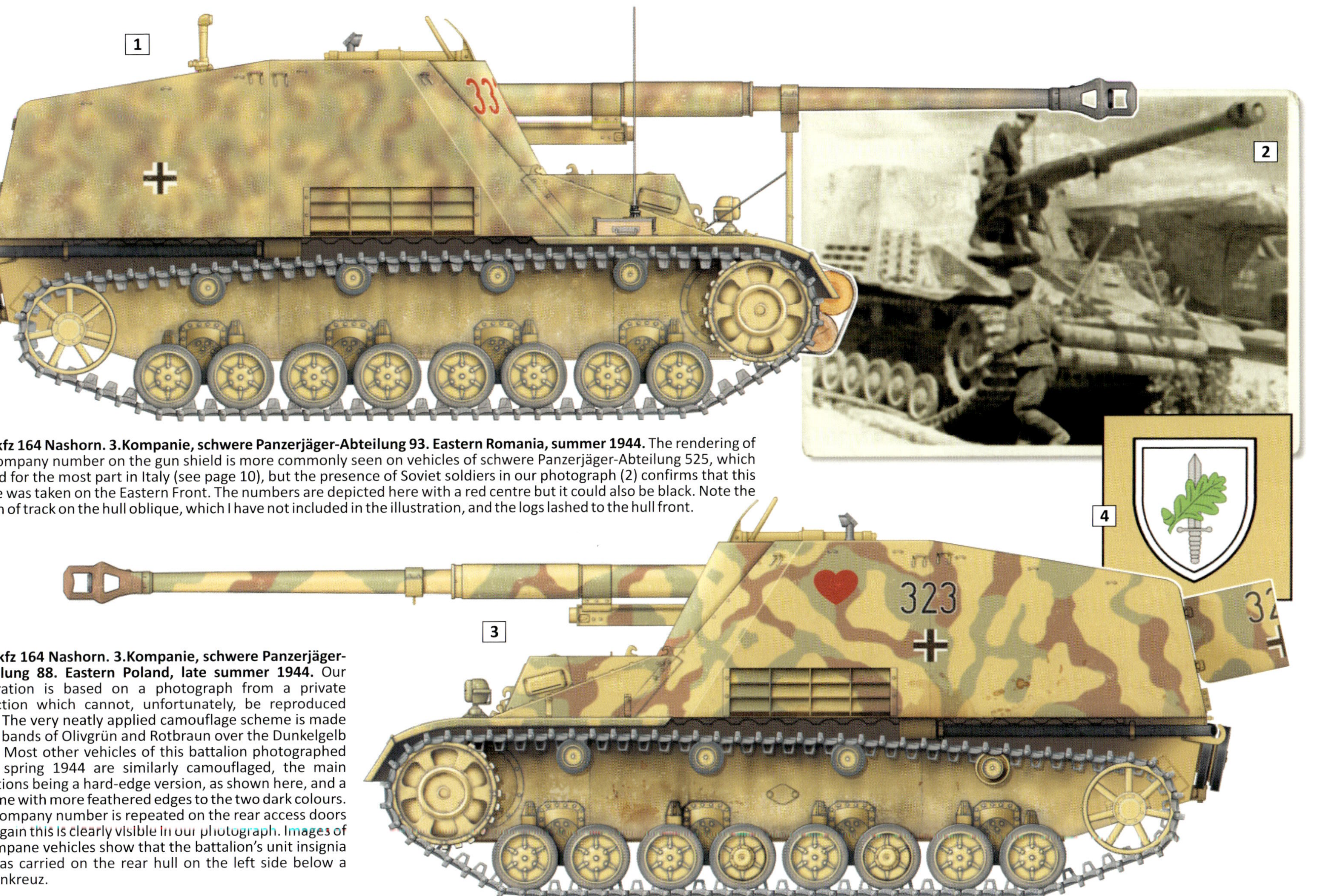

1. Sdkfz 164 Nashorn. 3.Kompanie, schwere Panzerjäger-Abteilung 93. Eastern Romania, summer 1944. The rendering of the company number on the gun shield is more commonly seen on vehicles of schwere Panzerjäger-Abteilung 525, which served for the most part in Italy (see page 10), but the presence of Soviet soldiers in our photograph (2) confirms that this image was taken on the Eastern Front. The numbers are depicted here with a red centre but it could also be black. Note the length of track on the hull oblique, which I have not included in the illustration, and the logs lashed to the hull front.

3. Sdkfz 164 Nashorn. 3.Kompanie, schwere Panzerjäger-Abteilung 88. Eastern Poland, late summer 1944. Our illustration is based on a photograph from a private collection which cannot, unfortunately, be reproduced here. The very neatly applied camouflage scheme is made up of bands of Olivgrün and Rotbraun over the Dunkelgelb base. Most other vehicles of this battalion photographed after spring 1944 are similarly camouflaged, the main variations being a hard-edge version, as shown here, and a scheme with more feathered edges to the two dark colours. The company number is repeated on the rear access doors and again this is clearly visible in our photograph. Images of 1.Kompane vehicles show that the battalion's unit insignia (4) was carried on the rear hull on the left side below a Balkenkreuz.

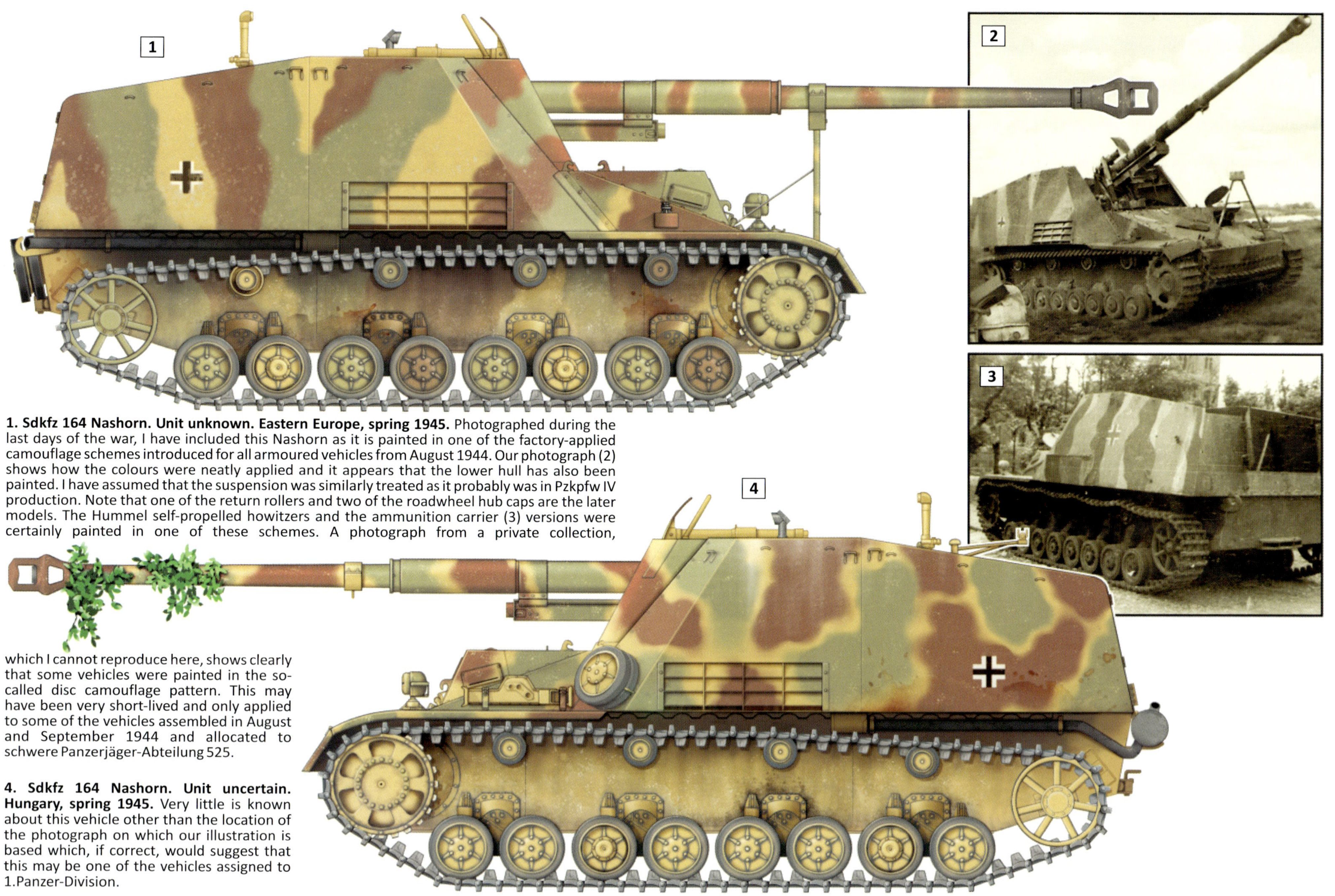

1. Sdkfz 164 Nashorn. Unit unknown. Eastern Europe, spring 1945. Photographed during the last days of the war, I have included this Nashorn as it is painted in one of the factory-applied camouflage schemes introduced for all armoured vehicles from August 1944. Our photograph (2) shows how the colours were neatly applied and it appears that the lower hull has also been painted. I have assumed that the suspension was similarly treated as it probably was in Pzkpfw IV production. Note that one of the return rollers and two of the roadwheel hub caps are the later models. The Hummel self-propelled howitzers and the ammunition carrier (3) versions were certainly painted in one of these schemes. A photograph from a private collection, which I cannot reproduce here, shows clearly that some vehicles were painted in the so-called disc camouflage pattern. This may have been very short-lived and only applied to some of the vehicles assembled in August and September 1944 and allocated to schwere Panzerjäger-Abteilung 525.

4. Sdkfz 164 Nashorn. Unit uncertain. Hungary, spring 1945. Very little is known about this vehicle other than the location of the photograph on which our illustration is based which, if correct, would suggest that this may be one of the vehicles assigned to 1.Panzer-Division.

NASHORN

LUC KLINKERS

1/35 SCALE

Luc's replica was based on the Dragon Models' 1/35 scale Nashorn, shown at left. The assembly was completed in sections to facilitate the detailing and painting of the superstructure interior and the gun behind the shield. Luc added details, many scratch-built, including the Fu 8 radio, towel and pistol holster, parts of the ammunition stowage lockers and the support for the gunner's seat. Note that mud has already been added to the hull sides.

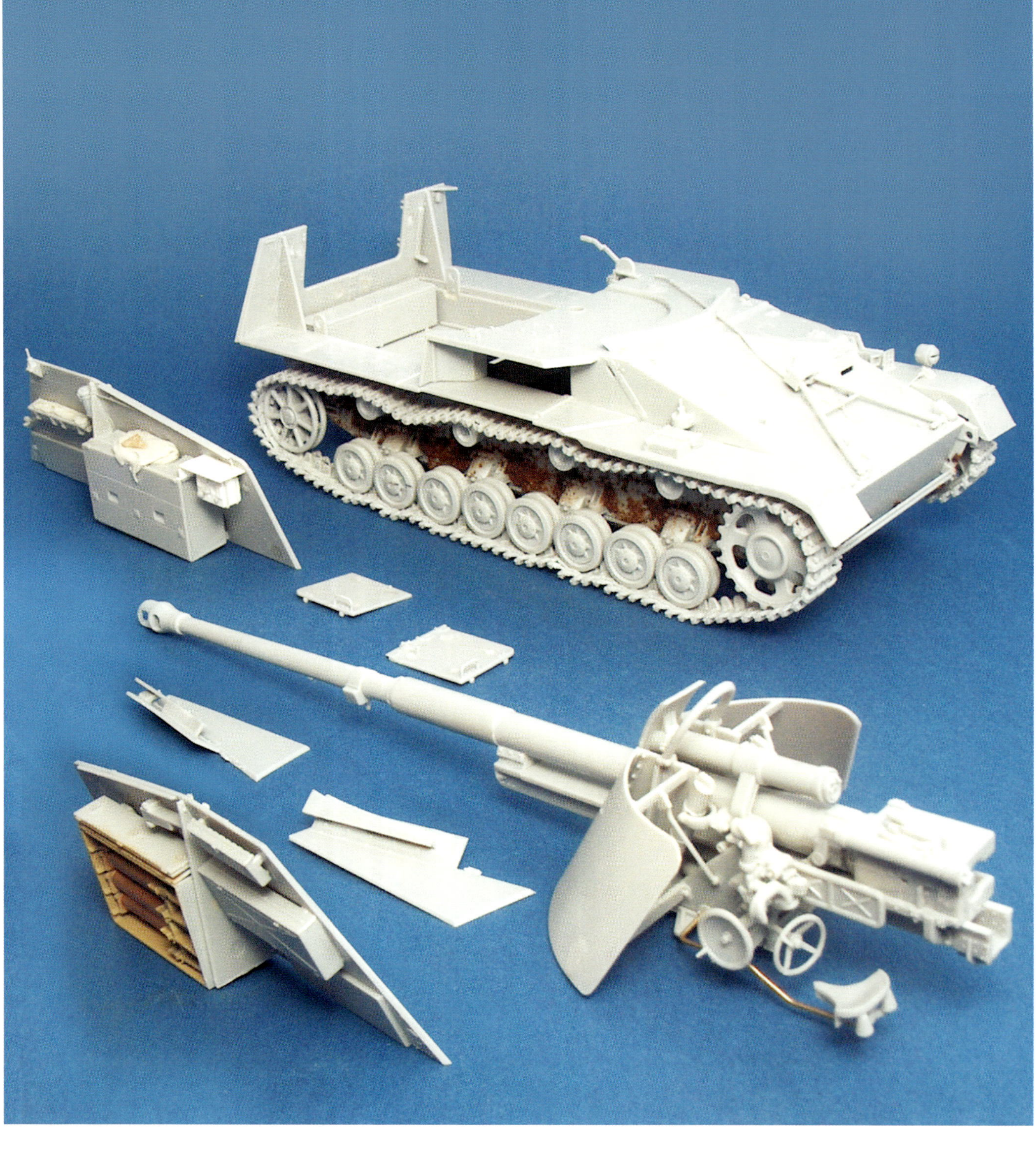

Above: The superstructure parts were temporarily held in place with masking tape to ensure that all the parts would fit correctly. Below: details of the partly assembled model including the ammunition stowage rack and an individual 8.8cm shell. This part was painted first and then fitted to the inside section of the superstructure side. Also shown below are the driver's visor, external gun travel lock, Bosch headlight and the lower hull with tracks fitted and mud texture added.

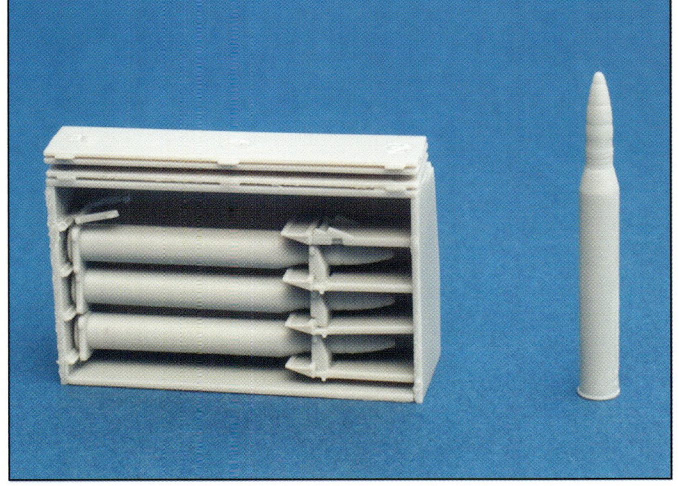

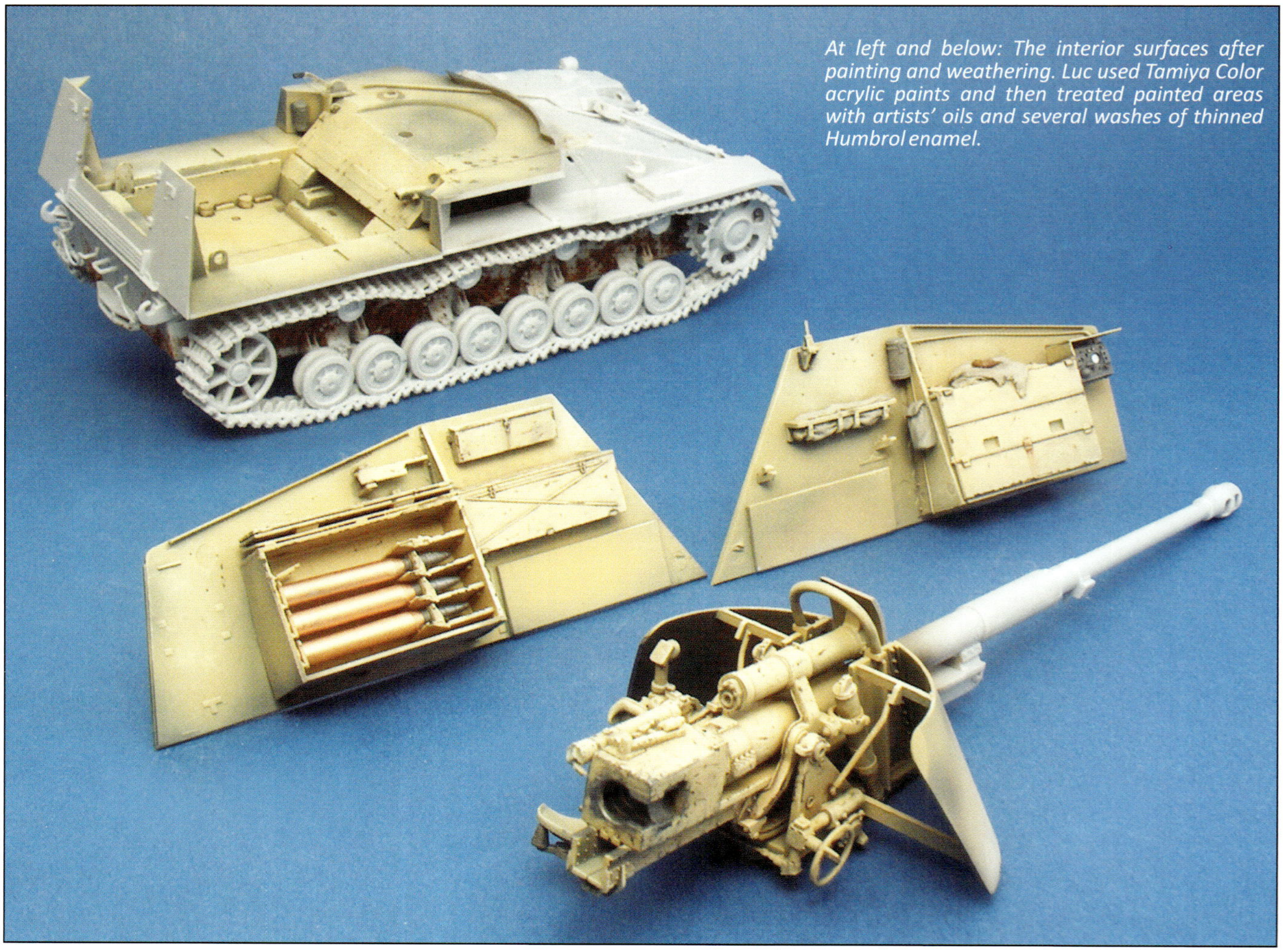

At left and below: The interior surfaces after painting and weathering. Luc used Tamiya Color acrylic paints and then treated painted areas with artists' oils and several washes of thinned Humbrol enamel.

The crew figures were created from Magic Sculpt putty built up over copper wire, using some parts from the kit. When the final details had been added Luc painted the figures using Humbrol enamel paints with artists' oils used for shading and texture.

The base was created from polystyrene foam, carved to shape and the coated in dirt. The finished product was painted in Tamiya Color acrylics.

The completed model on its base with the crew figures and accompanying infantryman in place. Note the extremely realistic barbed wire entanglements and parts of a Soviet T-34.

NASHORN

KRZYSZTOF KUDRYŃSKI

1/72 SCALE

Krzysztof's model is based on the Revell 1/72 scale kit, shown at left. For its size this is a detailed replica but Krzysztof has added small details such as brackets and hooks and some interior fittings. The finished model was painted with acrylic paints from the Vallejo Model Color range and weathered with subtle washes in artists' oils and products from the Ammo by MIG which also offers scenic details such as grass and trees. A prolific modeller, Krzysztof's work can be seen on his website Model 72 Miniatures where he uses small scale figures from Gebo Figuren, Redoak and Zvezda, many of which he converts for a specific project.

Above: Details of the completed model showing the infantryman and crew figures, hull rear with gun cleaning rods and spare roadwheels. Note the step between the two wheels, the gun breech and the right side air outlet. The hull has been treated with a fine chipping effect and dirt and dust is built up above the trackguards. Below: The completed model and figures on their base. Krzysztof's bases are constructed form a Balsa frame which supports the foam core.

NASHORN

M AHMET YAZAL

1/35 SCALE

Turkish modeller Ahmet's model is based on the Tamiya 1/35 scale Nashorn shown above. Below: The partly completed model, less the tracks, is shown here. Ahmet uses Tamiya Color acrylics and Tamiya Panel Line Accent, a very thin enamel-based product, to emphasise recessed areas. He also works with the Panzer Aces range of weathering colours from Vallejo and other weathering products from AK Interactive.

Above: The interior of the fighting compartment after weathering but prior to the extra small details being fitted. Note the driver's visor in the open position and the 8.8cm ammunition crate. The helmet and boots were taken from other kits but the jacket was sculpted from putty.

Above: A view of the completed model. The three-colour camouflage is similar to the schemes applied to the vehicles of schwere Panzerjäger-Abteilung 88 in the latter part of 1944 and is depicted in the illustration on page 27. At left: The rather cluttered, albeit very convincing, interior of the fighting compartment. Many of the details were taken from other kits, for example the ammunition crates, fuel can and helmets, but most of the others were scratch-built. Below: Another view of the completed model. The canvas cover for the superstructure is particularly realistic.

Model kits of the Hornissse/Nashorn self-propelled gun have been available since at least 1968 when Bandai released their 1/48 scale offering. Although somewhat rudimentary by today's standards this was an exceptionally detailed and accurate representation for its time and Bandai models are still eagerly sought. The number of kits on the market today is limited when compared to some of the more famous armoured vehicles but they are produced in a number of scales from 1/144 scale wargames tanks to the popular 1/35. They are vastly outnumbered by the upgrade sets which offer everything from detailed barrels to photo-etched interiors. For that reason I have limited the following list to complete kits, endeavouring to include some of the lesser-known brands, but I have included some of the producers of aftermarket sets in the index of manufacturers on page 64. They are listed here in no particular order and I have only commented on features of which I have personal experience.

AFV CLUB/HOBBY FAN

Hobby Fan Enterprises is the parent company of the better-known AFV Club, both of which are based in Taiwan. The current AFV Club catalogue includes a post-May 1943 production Nashorn in 1/35 first released in 2010. The kit contains plastic sprues, photo-etched details, a turned metal barrel, vinyl tracks and extensive markings in the form of waterslide transfers. AFV Club also offers an interior upgrade set and a detailed wheels and suspension set for the Nashorn or Hummel. An early production model is in the planning stage.

ZVEZDA

Founded in 1990, this Russian company has grown to become one of the country's larger exporters. In recent years Zvezda's catalogue has become heavily weighted towards modern Russian and Soviet-era vehicle and aircraft models but does include a 1/100 scale Nashorn which is part of a larger range of Second World War and Modern armour kits aimed at wargamers.

At left: The box of Zvezda's 1/100 scale Nashorn. Below: the completed model. The very basic detail is a feature of the models in this range but they designed to quickly, and rather cheaply, build wargames units. Given that, they are reasonably accurate, repay careful painting and look impressive when deployed en masse.

TAMIYA INCORPORATED

This Japanese company is currently the largest producer of scale model kits in the world and released their first 1/35 scale version of the Nashorn in 2014. This kit is made up of injection-moulded plastic sprues, in clear plastic for the optics, rubber parts, vinyl tracks, photo-etched details, anatomically correct crew figures and a sheet of high-quality waterslide transfers. In 2021 Tamiya released a 1/48 scale Nashorn model which is basically a scaled-down version of their 1/35 scale kit. Both models depict later production versions.

At left: Tamiya's 1/35 scale Nashorn built with the photo-etched brass detail set from Eduard Model accessories. Below: Details of the 1/35 scale kit built and painted from the company's catalogue and below that, details of the 1/48 scale model.

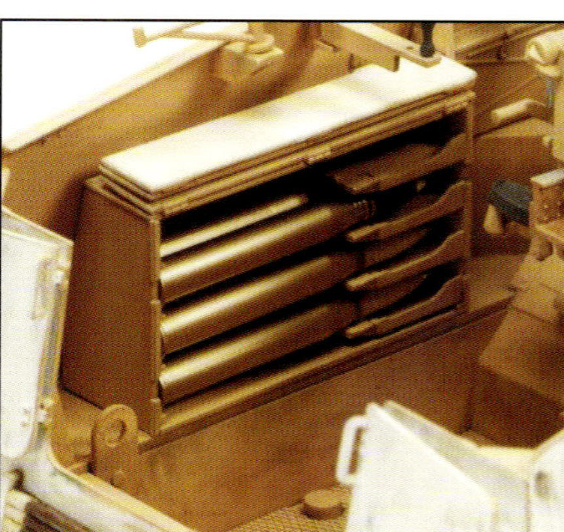

BORDER MODEL

Situated just outside Shanghai, this company produces a large range of armour and aircraft kits. The 1/35 scale Nashorn is marketed as giving the option to build an early or command version but in fact contains part for a version with the cylindrical exhaust, the simplified gun travel lock or the initial variant with its internal support. The tracks are injection-moulded plastic, made up of individual links, and a small photo-etched brass set is included. The instructions are quite detailed and include photographs of the real thing and some very attractive colour profiles.

At right: The box for Border Model's 1/35 scale Nashorn with artwork by the very talented Jason Wong. Below: The 1/35 scale model completed with the parts supplied with the kit. Below, left: The four crew figures.

REVELL

This company has gone through many changes of ownership over the years since its founding in 1943, at one time operating as two distinct entities in the US and Germany. The company's current catalogue includes a 1/72 scale Nashorn which was first released in 2005. In cooperation with IPMS Deutschland a limited-edition version was also released which included new parts.

Revell's 1/72 scale Nashorn. Markings for a vehicle of schwere Panzerjäger-Abteilung 88 are also included with the kit.

DRAGON MODELS

Based in Hong Kong, Dragon Models was one of the first of the major manufacturers to offer accurate replicas in 1/35 scale and their kits are today something of a benchmark. The company's small scale releases are probably the best available. In 1/35 scale the company offers a number of kits including early and later production variants which are essentially based on the 2002 release which has been continually upgraded with new parts and improved tooling. These are complemented by a 1/72 scale range which first appeared in 2002. Dragon also produces 1/72 assembled and pre-painted models of the Nashorn.

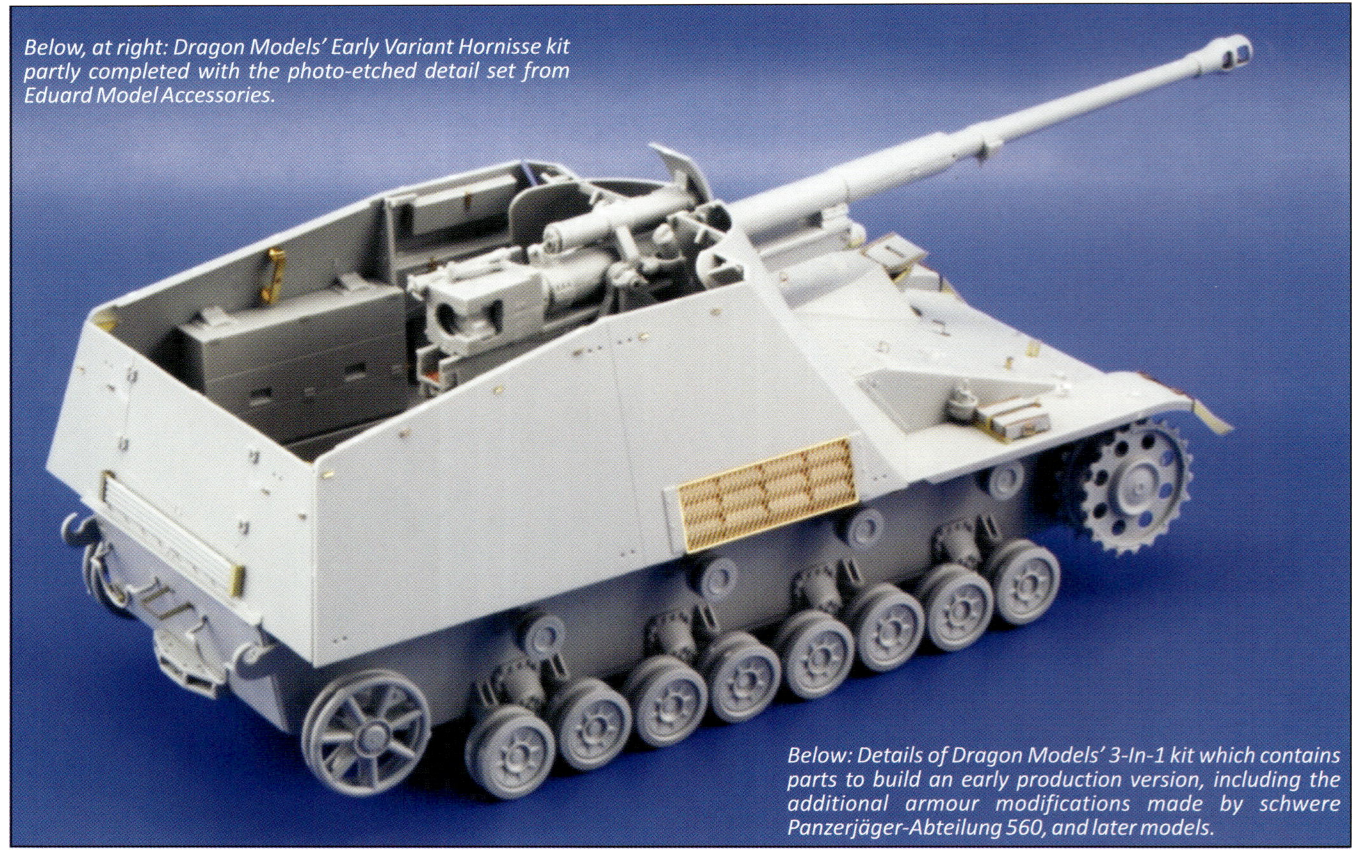

Below, at right: Dragon Models' Early Variant Hornisse kit partly completed with the photo-etched detail set from Eduard Model Accessories.

Below: Details of Dragon Models' 3-In-1 kit which contains parts to build an early production version, including the additional armour modifications made by schwere Panzerjäger-Abteilung 560, and later models.

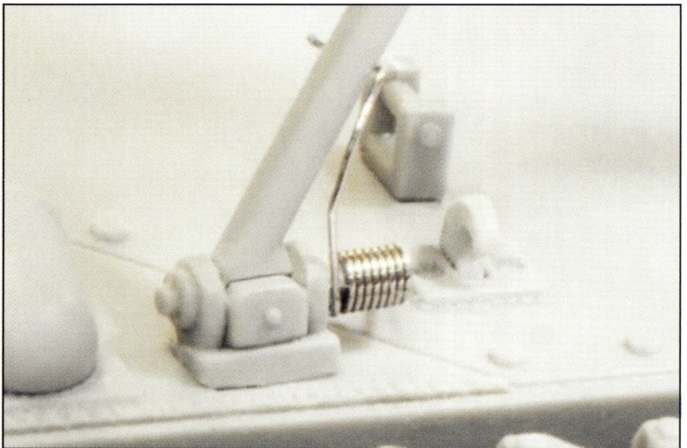

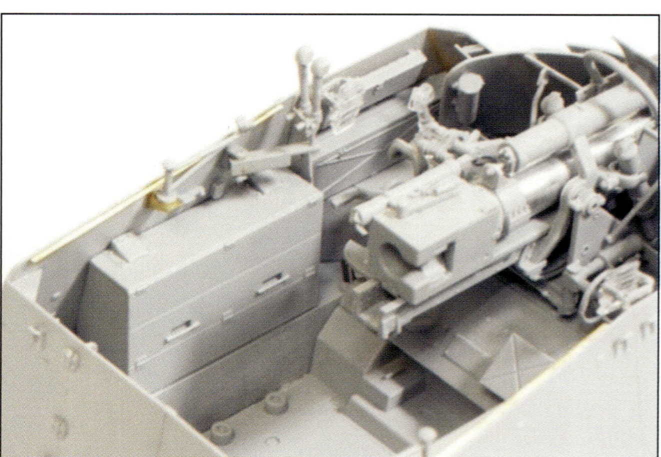

BANDAI

In the late 1960s and early 1970s this Japanese company produced an extensive range of Second World War armour models in 1/48 scale. The series was popular but with the entry of Tamiya into the market with their 1/35 scale offerings, which were in some cases in the same price range, sales fell off and Bandai eventually abandoned production of armour kits entirely. Frog and Fuman released several of the Bandai kits in the 1990s but today they are only available through specialist dealers or on-line auctions. The Nashorn kit, marketed as Anti-Tank Gun IV 'Nashorn', featured an interior where the detail was mostly moulded-in and would be considered as very basic by modern standards. In spite of all this, Bandai kits have largely retained their popularity, especially with Japanese modellers.

At left: Bandai's 1/48 scale Nashorn showing the original packaging, the marking options and a kit part comprising the hull front and fighting compartment floor. Note that a great deal of the detail is moulded in.

BOX STUDIO

This Japanese company offers a range of paper models covering a number of well-known armoured vehicles, and some rather more obscure, in 1/48 and 1/72 scale. The Nashorn in 1/48 scale is shown below. As far as I can determine, there is no English-language version of the website, the models can be downloaded as PDF files for free.

Above: Box Studio's 1/35 scale Nashorn made entirely from paper. The paint scheme is already printed on the model parts but could of course be altered.

ROCO MINITANKS

Founded in 1960, this Austrian company produced a large range of armoured vehicles in 1/87, or HO scale. This was a popular size in Europe at the time as it complemented many model railway products. In 2005 the company went into bankruptcy and two years later the Minitank moulds were acquired by Herpa Miniaturmodelle, a German firm which specialised in die-cast model cars and aircraft. The Nashorn model is currently listed in the company's archive but is re-released from time to time.

Roco Minitanks Nashorn comes fully assembled as shown here. Some details are finely rendered while others are basic and even inaccurate, such as the Pzkpfw IV drive sprocket.

PLANET MODELS

Based in the Czech Republic, this company specialises in resin armour and aircraft kits. The 1/72 Nashorn was first released in 2002 and includes a turned metal barrel, photo-etched details, vinyl tracks and waterslide transfers. This kit was originally offered by Modellbau Schatton and was probably based on that company's Hummel kit

BOLT ACTION

This brand is actually part of the Warlord Games franchise which itself started life as a two-man operation in 2007 and has grown into an international concern. Bolt Action figures and vehicles are produced in 28mm scale, which equates to 1/56, and the Nashorn (shown below) is made up of resin and metal parts and contains four crewmen.

As mentioned in the introduction to this book, the development of the Nashorn tank destroyer can be traced to June 1942 when Hitler inspected the Rheinmetall-Borsig 8.8cm Flak 41 anti-aircraft gun and demanded that an anti-tank weapon with comparable range and the ability to penetrate the thickest enemy armour be built. Both Rheinmetall-Borsig and Friedrich Krupp AG were almost immediately contracted to produce different models. It is probable that the decision to produce towed and self-propelled versions was taken shortly after this as Reichs Armaments Minister Speer discussed with Hitler in late July 1942 the plan to mount the prototypes of each design on the le FH 43 (sfl) which was then under development at Skoda. The request for a self-propelled mount may in fact have been contemplated at the same time by the officials of the Heereswaffenamt (HWA) who were well aware of the paucity of vehicles capable of towing such a heavy weapon. Consideration was also given, albeit briefly, to mounting the 8.8cm gun on the chassis of the planned Geschützwagen II, as eventually used for the Sdkfz 124 Wespe.

At the same time the management of Rheinmetall-Borsig requested clarification on which type of ammunition was to be used with the new gun as they had learnt that the Krupp model was being developed to fire the company's Flak 42 Patrone, initially designed for the 8.8cm KwK L/71. The cartridge of the Krupp round was of slightly larger diameter than Rheinmetall-Borsig's 8.8cm Flak 41 Patrone, which had already been tested and accepted by the HWA, and although it would allow for a more powerful charge it would significantly decrease the lifespan of the gun tube. In addition, the Rheinmetall-Borsig design incorporated the type of breech that had been proposed for the 7.5cm Pak 42 of the Skoda le FH 43 (sfl) programme and the roller loader that had already been accepted. Neither of these features were compatible with the Krupp 8.8cm round. The HWA may have initially agreed with this, or at least prevaricated, but by 5 September 1942 a decision was made to replace the Rheinmetall-Borsig cartridge with the Krupp design. To ease production, the barrel was to be constructed in two parts and with an overall length of 6380mm, much shorter than the 8.8cm Flak 41. On 11 September representatives of both firms were advised that a final decision had been made that the gun tubes of what was now referred to as the Pak 43 would be employed with the carriage of the 10cm Kanone, necessitating an almost complete redesign. Drawings were to be completed by the end of October 1942 and a trial version ready for test firing by 20 December. Series production would be undertaken by Henschel & Sohn and it was expected that 500 of these guns would be completed by May 1943.

The option of using the le FH 43 (sfl) as a suitable platform was very quickly reconsidered when it was realised that the programme would not come to fruition by early 1943 and on 28 July 1942 it was decided that a Zwischenlösung or expedient design, should be produced as soon as possible using available components. The firm of Altmärkische Kettenfabrik (Alkett) was awarded the contract and, guided by the HWA, produced a vehicle based on elements of the Pzkpfw III and Pzkpfw IV tanks and referred to as Geschützwagen (GW) III/IV. On 2 October 1942, Speer recorded that the Alkett design, which could accommodate both the 8.8cm anti-tank gun and the 15cm howitzer, was ready and that plans were made to complete 100 of each version by May 1943. Interestingly, all the design and development work had been carried out on the Hummel 15cm-armed variant and no prototype of 8.8cm-equipped vehicle, by now christened Hornisse or hornet, was ever built.

As the Hornisse/Nashorn was produced in relatively small numbers, with just 494 vehicles leaving the assembly lines before production was terminated, few production modifications were made to the basic vehicle. For our purposes it is reasonably accurate to divide the completed vehicles into two broad categories: early or initial production versions, which were assembled from February 1943, and a standard variant which appeared between May and August of the same year. The known modifications are listed below but it should be emphasised that the edges here could easily be blurred and early production features are commonly seen on vehicles known to have been built in August or later. Similarly, thirty-seven vehicles were, over time, returned to Germany and we have no way of knowing which, if any, modifications were incorporated into their repair. A number of modifications were made by operational units in the field and these are shown and described in the accompanying photographs as these images are the only evidence we have.

January 1943. Alkett was awarded a contract on 22 January to assemble a total of 420 Nashorn (1). The first ten assemblies were to be completed before the end of the month with a further twenty to be delivered in February. Production was planned to rise to thirty vehicles per month by March 1943 and this rate was to be maintained until the end of March 1944 when production would cease. On the same day, Deutsche Eisenwerke AG was contracted to build a further 150 vehicles at the company's Werk Stahlindustrie plant at Duisburg and it was anticipated that the first five would be completed in May 1943. Production was planned to rise to fifteen per month by the following July and end in March 1944 (2). The basic armoured hulls for both the Nashorn and Hummel programmes were constructed by Witkowitzer Eisenwerke.

1) Hull and superstructure.

The Panzerwanne, or armoured hull, was constructed of rolled carbon steel (SM-Stahl) plates of 10mm thickness, except the rear plate which was 20mm thick, and the front plate which was 30mm face-hardened armour plate. Two access doors, which opened outwards, were located on the rear plate of the superstructure.

The hull sides were extended at the front to create towing eyes and the armoured covers for the brake vents situated on the hull glacis were identical to those fitted to the Pzkpfw III ausf F. The driver's compartment was located in the hull front and was fully enclosed. The armoured superstructure which protected the main gun and the crew was also built from

Notes

1. Once again, simply as a matter of convenience, I have referred to the 8.8cm self-propelled gun throught this section as Nashorn except where noted.

2. These projections, as will be seen, were almost constantly revised throughout the vehicle's production run.

Hitler, at centre with Generalfeldmarschall Keitel on his right, photographed in January 1943 at a presentation of new weapons systems and prototypes. Reichs Armaments Minister Albert Speer is standing at the base of the 8.8cm rounds, next to the civilian gentleman with the dark hat. In the background is an 8.8cm Pak 43, the first version built on a cruciform mounting, and behind that one of the first production Nashorn self-propelled guns. Note that the latter is already painted in what appears to be a three-colour camouflage scheme. Behind the Nashorn is an Sdkfz 124 Wespe and, just visible, the barrel of an Sdkfz 165 Hummel.

Notes

1. Most of the vehicles that can be observed in photographs which have spare road wheels mounted on some part of the forward hull or superstructure are fitted with the early gun travel lock.

10mm-thick rolled carbon steel. The original requirement that the hull and superstructure be composed of 20mm carbon steel with a 50mm-thick front plate was altered to save weight. The superstructure was open-topped and the crew were only protected from the elements by a removable canvas cover. The hull glacis was sloped at 70 degrees except for the raised driver's position which featured vision ports on either side, a visor at the front and a large round hatch on the roof which opened forwards. Another similar hatch was located on the right side of the glacis next to the driver's position above the radio operator.

Two Bosch headlights were mounted at the front on each trackguard and the jack was placed on the left side trackguard and the wooden jack block was on the right side.

Spare roadwheels were held in brackets of varying design and the placement of the brackets and wheels varied. On Fgst Nr.310002, assembled in February 1943, and Fgst Nr.310030, built in the following March, the wheels are fitted on the superstructure oblique, facing forward. On Fgst Nr.310042, which left the assembly plant in late March, one wheel was held on the right side oblique and one on the left side superstructure front. Combinations of these placements can be seen in contemporary photographs but unfortunately the individual vehicles cannot be identified. The practice seems to have ended at some time after April 1943 when the exhaust muffler was dropped from production and standard brackets for spare roadwheels were welded to the hull rear plate (1). It has been argued that these were field modifications but similar arrangements seen on Hummel self-propelled howitzers would suggest that many were applied in the factory.

2) Suspension.

The series production vehicle was designed with the same width as the Pzkpfw III, which was wider than the Pzkpfw IV, and this meant that the drive train components including the final drives, steering units, drive shafts and transmission, were adopted from the former.

The cast drive sprocket was taken directly from Pzkpfw III ausf H production. On each side of the hull eight sets of 470mm-diameter roadwheels were mounted in pairs on four elliptic leaf-spring bogies and the track was supported by four return rollers. All these components and the rear idler were identical to those fitted to the Pzkpfw IV ausf F. At some point during

..........text continued on page 52

Photographed in Germany in early 1943 this Nashorn displays several identifying features of early production models including the Pzkpfw III ausf H drive sprocket, the first pattern external gun travel lock, twin Bosch headlights and the armoured box covering the exhaust pipe at the hull rear. The officer standing behind the rear access doors may be supporting himself on steps similar to those welded to the vehicle above the cylindrical exhaust muffler. This feature is clearly visible in the photographs on pages 53 and 54.

Assembled in February or March 1943 this Nashorn was allocated to schwere Panzerjäger-Abteilung 560. Note the first pattern gun travel lock and twin Bosch headlights. The antenna insulator for the Funksprechgerät f transceiver can be seen behind the small jack block fitted to the trackguard. The placement of the spare roadwheel on the superstructure oblique seems to have been common with vehicles of this battalion. Other photographs of this vehicle show that it was fitted with the Pzkpfw III ausf H drive sprocket.

A number of accounts suggest that this Nashorn may be Fgst Nr.310001, the first example of the series, but I have been unable to confirm this. In any case it is a very early production vehicle and may be the same Hornisse shown on page 50. Note the first pattern external gun travel lock (1) and support (2), cast drive sprocket (3), hull extensions (4), the position of the towing cables (5), the metal cover for the ZE 34 indirect-fire sight aperture (6), twin Bosch headlights (7) and jack (8). Note also that there is no provision for the Fuchs-Gerät coolant heater which was introduced into production in March 1943.

..........text continued from page 50

series production, it is unclear when, the distance between the rear set of road wheels and the idler was increased. The production series Nashorn were fitted with 400mm-wide Kgs 61/400/120 tracks.

3. Engine and exhaust system.

The Maybach HL 120 TRM engine, cooling fans, radiator and exhaust muffler were taken from the Pzkpfw IV ausf F. The engine was moved from the rear to the centre of the hull and its covering deck provided a platform for the gun mount which was closer to the vehicle's centre of gravity.

Vents on the left side of the chassis pulled cooling air through the radiator which then passed over the engine and was emitted by fans through the vents on the right side.

Two exhaust pipes ran from the engine along each side of the hull and were fed into a cylindrical exhaust muffler which was fitted to the hull rear below the access doors. The exhaust was emitted through a pipe on the top of the muffler and this proved to be problematic as the fumes tended to rise and accumulate in the fighting compartment.

In an effort to combat this, extensions were welded to the pipe and at least five different arrangements can be observed in contemporary photographs but it is unclear if these were factory modifications or expedients effected by units in the field. Given that the exhaust mufflers were only fitted between February and April 1943, effecting eighty-five vehicles at most, the latter would seem to be very likely.

4. Gun mounting.

The 8.8cm Pak 43/1 (L/71) gun, as mounted in the Nashorn, was identical to the towed version including the upper carriage. The shape of the gun shield was altered to better fit the superstructure sides as the gun was traversed. The recuperator was mounted above the gun tube and the recoil cylinder was fitted below the barrel and two counterbalance cylinders were mounted of each side of the gun.

The first examples featured an external travel lock, as designed for the Hummel, and an internal lock. The gun was provided with a direct-fire sight to be used with armour-piercing rounds and an indirect sight for high explosive shells (see April 1943). Two ammunition bins which each held eight 8.8cm rounds were located on either side of the crew compartment interior. Vehicle crews were authorised to carry twenty-four additional rounds in wooden crates on the crew compartment floor.

5. Radios.

Each vehicle was equipped with a Funksprechgerät f (1) transceiver with either a 1.4-metre or 2-metre whip antenna. This radio provided voice-based communication with other vehicles and had a maximum range of 5 kilometres under optimum conditions. The equipment and its operator were located in the forward hull next to the driver and the antenna and insulator were mounted on the right side of the superstructure behind the wooden jack block. Vehicles assigned to battalion and company headquarters were also fitted with the Funkgerät (Fu) 8 radio, a medium-wave transceiver that

..........text continued on page 54

Notes

1. This radio was commonly fitted to open-topped vehicles.

Fgst Nr.30030 assigned to 1.Kompanie, schwere Panzerjäger-Abteilung 560 photographed shortly after its capture by the Soviets. The gun travel lock has been removed but otherwise this vehicle is in reasonable condition. Note the additional armour (1) on the gun shield and the rain guard above the radio operator's hatch (2) which were field-modifications unique to this battalion. Note also the twin Bosch headlights (3), early-pattern brake vent covers (4) and the position of the towing cable brackets (5) and spare roadwheel (5). This vehicle is currently on display at the Kubinka Tank Museum, Moscow.

The same vehicle photographed from above. Note the armoured box (1) covering the exhaust pipe and the mudguard (2) fitted to the very earliest vehicles. As with most vehicles the outlet on the top of the cylindrical exhaust has been covered (3) and an extension fitted to the left side (4). The covers above the exhaust (5), which also served as additional steps, the handle on the superstructure rear (6) and bracket on the right side rear access door (7) are all field modifications.

A very early production Hornisse fitted with the cylindrical exhaust muffler. Note that the original exhaust outlet (1) has been covered and welded shut and a pipe has been added to the left side (2). One of the armoured boxes (3) is still in place as are the ranging stakes (4). The gun shield lacks the strengthening metal loop that was incorporated into production in May 1943 and the Zieleinrichtung 43 direct-fire sight (5) is still in use. Note the position of the steps on the hull rear (6) and the covers above the muffler (7) which also served as steps. The latter is a field modification that seems to have been peculiar to schwere Panzerjäger-Abteilung 560.

..........text continued from page 52

Notes

1. A combat report compiled by the commander of sPzJg-Abt 560 requested that two extra radio operators be authorised for each command vehicle but this suggestion was apparently never acted upon.
2. This was actually a trade name.
3. Artillery ranging stakes or poles, painted red and white, were used in the process of laying and firing howitzers, in this case the 15cm gun.

was used to communicate with higher command. It had a range of 25 kilometres when used with the 1.8-metre antenna. The radio, antenna and insulator were mounted at the rear of the crew compartment and an obvious drawback was that the radio operator needed to leave his position to access the Fu 8 and this problem was never solved (1).

February 1943. During the first week of this month Hitler directed that production of the Hummel 15cm self-propelled gun was to be prioritised and Deutsche Eisenwerke was directed to abandon the Nashorn project. As reported by Reich Minister Speer, planned production of the Nashorn was revised from forty-five to twenty vehicles per month. The first production examples, beginning with Fgst Nr.310001, left the assembly line at Alkett and by the end of the month a total of fourteen vehicles had been accepted by the HWA. Towards the end of the month, a curved armoured shield was fitted to the recuperator cylinder.

March 1943. The armoured covers for the brake vents on the hull glacis, identical to those fitted to the Pzkpfw III ausf F, were replaced by larger, more rounded versions similar to those fitted to the Pzkpfw IV. The position of the towing cable brackets on the hull glacis was altered and a U-shaped bracket was added which secured both ends. A hole was cut into the left side of the hull above the roadwheel bogies to accommodate a coolant heater referred to as a Fuchs-Gerät (2) which allowed the engine to be preheated by a blowtorch. The hole was covered by an armoured cap.

Ranging stakes, as fitted to the Hummel on the superstructure rear below the access doors, were dropped from production (3). The brackets for the gun cleaning rods were retained but were now welded closer together.

In this month the initial production drive sprocket, identical to that fitted to the Pzkpfw III ausf H, was replaced by the version designed for the Pzkpfw III ausf E, which was flat and featured eight evenly-spaced holes.

It should be noted that this modification was only carried out intermittently and it is not clear when the cast drive sprocket of the Pzkpfw III ausf H, which had been fitted to the February production vehicles, became standard. Both are seen in photographs of what could be considered late production Nashorn. An OKH order of

22 March authorised the formation of the first battalion, sPzJg-Abt 560, and the vehicle was referred to as 's-Pak(sfl) Hornisse'. A total of thirty vehicles were assembled in March.

April 1943. The first fifty 8.8cm 43/1 guns were fitted with both the Zieleinrichtung (ZE) 43 mit Zielfernrohr 3x8 direct-fire sight and the ZE 34 mit Rundblickfernrohr (Rbl.f) 32 or 36 which was used for indirect fire but in this month they were both replaced by the ZE 37/43 with the periscopic Sfl.ZF.1a sight.

The adoption of this universal system may have been an effort to simplify production and negate the need to train the gunners on two different systems. It is often mentioned that the earlier sights were easily misaligned when the gun was fired or simply by the movement of the vehicle but reports to this effect do not seem to have originated before May.

The exhaust muffler was no longer fitted and in its place brackets were welded to the rear hull to hold two spare roadwheels. The exhaust pipes were shortened and towards the rear were angled outwards. Two tow hooks were now welded to the hull. The internal lifting eye, which had been bolted on to the inside of the hull rear, was moved forward and welded in place on the hull side. The armoured boxes mounted on each side at the rear below the hull panniers were dropped from production (1).

Alkett reported that forty-one vehicles were completed in April. An OKH order dated 22 April authorising the formation of sPzJg-Abt 525 referred to the 8.8cm self-propelled gun as '8,8cm Pz.Jag. IV b'.

May 1943. The headlight on the right side trackguard, next to the radio operator's position, was dropped from production in this month. As the parts had already been fabricated, the armoured cover for the electrical lead could still be seen on vehicles assembled in August. An improved external gun travel lock, which could be disengaged from the driver's position, was introduced into production. The version fitted to earlier vehicles could only be lowered by hand which necessitated the gunner leaving the protection of the superstructure.

The internal travel lock, mounted inside the superstructure behind the gun breech, was no longer needed and this was dropped from production. A metal strengthening loop was bolted to the inner surfaces of the gun shield over the recuperator cylinder and the thickness of the shield was increased from 10mm to 15mm. A HWA report dated 11 May 1943 stated that from the following September the 8.8cm Pak 43/1 was to be replaced in Nashorn production by the Pak 43. The vehicles were to be completed at a rate of forty per month from October 1943 to February 1944 when production would cease. A total of thirty-five vehicles were assembled in May.

..........text continued on page 57

Notes

1. The true purpose of these boxes is unclear and they are generally considered to have provided some kind of protection to the exhaust pipe.

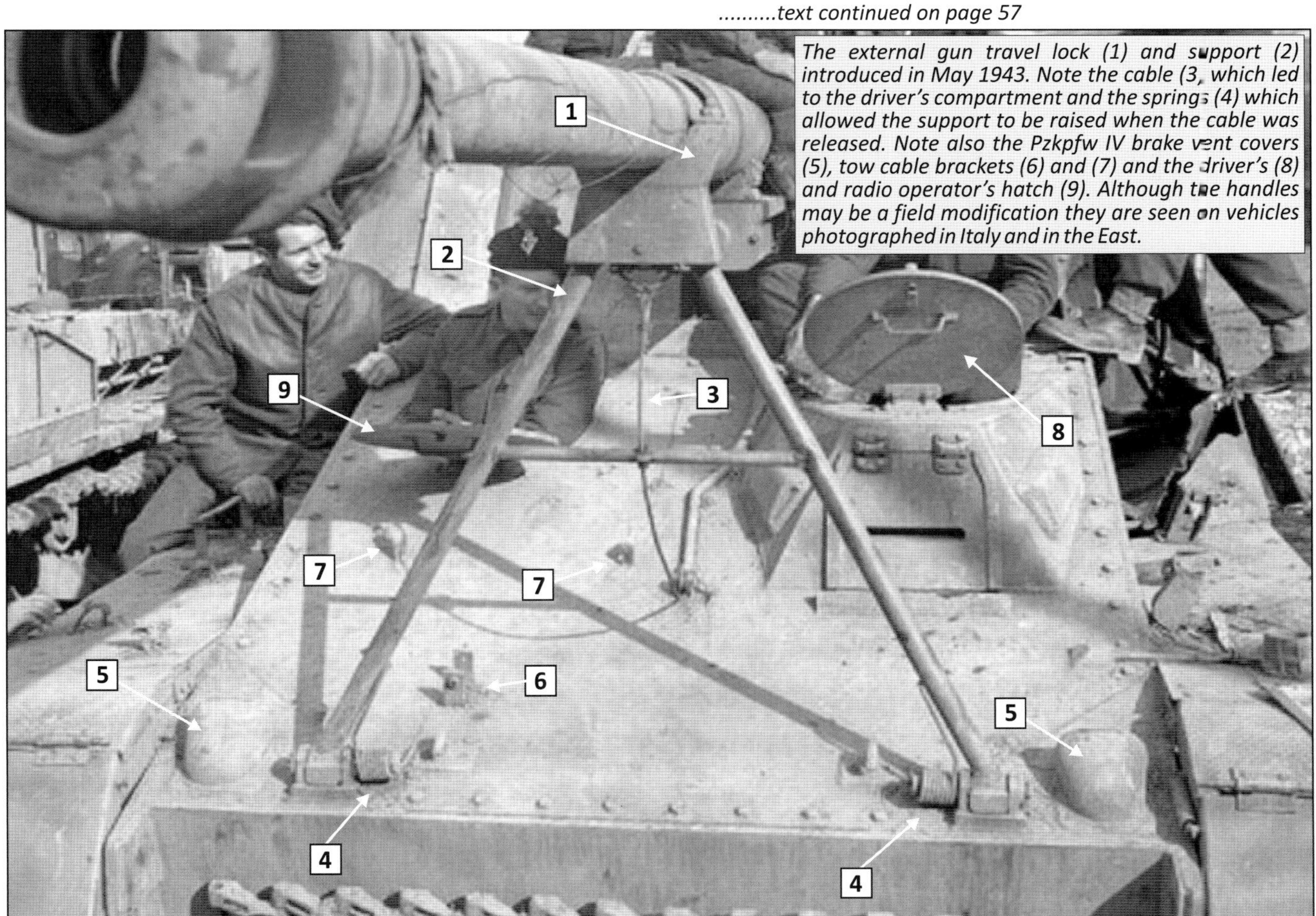

The external gun travel lock (1) and support (2) introduced in May 1943. Note the cable (3, which led to the driver's compartment and the springs (4) which allowed the support to be raised when the cable was released. Note also the Pzkpfw IV brake vent covers (5), tow cable brackets (6) and (7) and the driver's (8) and radio operator's hatch (9). Although the handles may be a field modification they are seen on vehicles photographed in Italy and in the East.

The images on this page both depict vehicles assembled during or after April 1943 when the cylindrical exhaust was dropped from production but both exhibit a mix of old and new features. Points to note are: 1. The first-pattern external gun travel lock which was phased out from May 1943. Although it is difficult to be certain, this Nashorn appears to lack the curved shield for the 8.8cm gun which was introduced in late February. 2. Canvas foul-weather cover. 3. Pzkpfw III ausf E drive sprocket which partially replaced the Pzkpfw III ausf H type. 4. Re-designed, outward-angled exhaust pipe, the end of which is just visible here. 5. Tow hooks welded to the hull side. 6. Track pin return plate. 7. Spare roadwheel brackets. 8. Crew step. 9. Gun cleaning rod brackets. Below: 10. The internal gun travel lock with its adjusting wheel (11). 12. The gun cleaning rods in place. 13. The exhaust pipes. Notice that these are much longer than the versions shown above and may in fact be cut-down versions of the first type shown on page 54. 14. The welded-on tow hooks showing how they are angled inwards. 15. The spare road wheel in its bracket. 16. The step which the crew used to enter the superstructure.

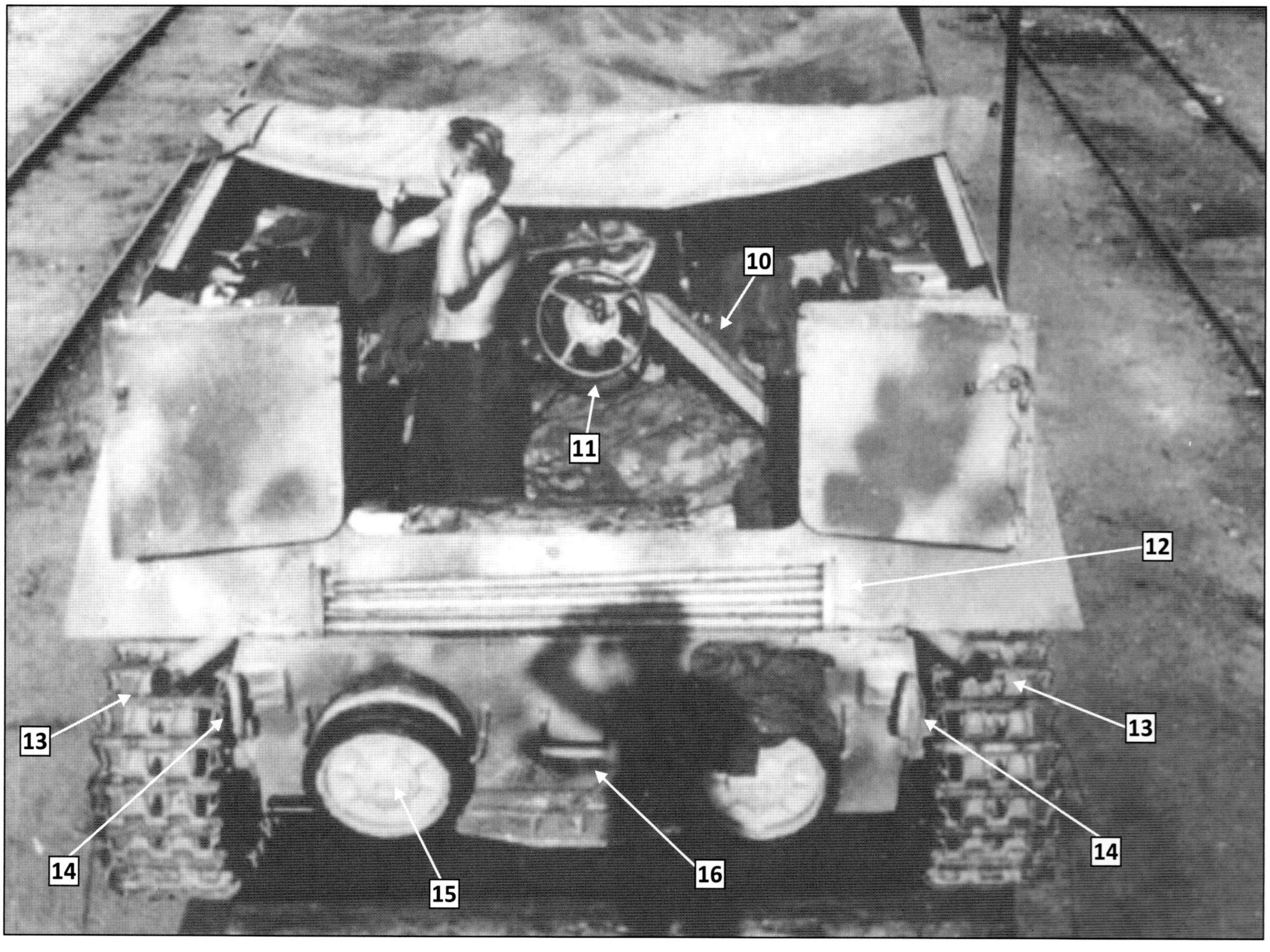

..........text continued from page 55

June 1943. A total of thirty-five vehicles were assembled in June and by the end of the month 105 Hornisse were with operational units.

July 1943. Plans to install the 8.8cm Pak 43 in place of the Pak 43/1 were abandoned and production of the latter was to be increased to sixty per month. Alkett reported that forty-four vehicles were completed in this month.

August 1943. The towing cable brackets on the hull glacis were relocated farther to the right, as viewed from the front, and the U-shaped brace was fitted lower than the z-shaped brackets. Allied bombing meant that just twenty 8.8cm Pak 43/1 guns were completed by Henschel in this month and just sixteen completed vehicles left the assembly lines at Alkett.

September 1943. Twenty-seven vehicles were completed in this month

October 1943. Forty-two vehicles were completed. In this month cast steel return rollers, which replaced the rubber-tyred versions, and roadwheels with cast hubcaps were introduced into Pzkpfw IV production. A small number of photographs exist of vehicles fitted with these new items but it would seem that their adoption was not universal.

November 1943. A combat report dated 21 November stated that the sight of the 8.8cm Pak 43/1 gun was easily prone to misalignment which could be exacerbated simply by driving the vehicle. Although it is not mentioned specifically, this must have concerned the Sfl.ZF.1a sight as all the earlier versions had been replaced by this time. The report went on to suggested that production of the Nashorn be terminated but a request was made by Insp.d PzTr that production continue until at least February 1944, when the Jagdpanther was expected to enter service. But within days the Alkett factory was devastated by an Allied bombing raid and although production was moved to another location, just twenty-four vehicles were completed in November, slightly more than half the number planned.

December 1943. Although the heavy bombing raids of late November were still affecting production a total of thirty-seven Nashorns were assembled in December, just three short of the contracted target. By the end of the year Witkowitzer Eisenwerke had completed all the hulls required for Nashorn production and this almost certainly accounts for the absence of modifications which were carried out in Hummel and Pzkpfw IV production after this time.

January 1944. No vehicles were built during this month and it must have been at this time, or perhaps slightly earlier, that the decision was made to give priority at the Alkett plant to the assembly of the Sturmgeschütz III and that Nashorn production would restart at Deutsche Eisenwerke.

February 1944. An OKW memorandum dated 1 February advised that the '8,8cm Pak auf Fgst. III/IV' was henceforth to be referred to as Nashorn (1). Production ended at Alkett with Fgst Nr.310370 giving a total of twenty-five vehicles assembled in this month. The management at Deutsche Eisenwerke reported that twenty-five 8.8cm Pak 43/1 guns and their upper carriages had arrived at the company's Teplitz-Schönau plant in the present-day Czech Republic.

Notes

1. The memorandum began with the sentence 'Der Führer hat nachstehende Suggestivnamen für Heer und Luftwaffe', meaning that the names originated with Hitler. Most accounts suggest that Hornisse was felt to be insufficiently aggressive but the Messerschmitt Me 410 fighter retained the title.

Although this vehicle is heavily damaged I have included this photograph as it is one of the very few images which clearly show the cast roadwheel hubcaps and all-metal return rollers fitted to a Nashorn. Both were introduced into the production process of the Pzkpfw IV in about October 1943 and it may be safe to assume that some found their way into the Nashorn programme at the same time.

Two Nashorn self-propelled guns probably photographed in early 1944. The vehicle closest to the camera is fitted with the Pzkpfw III ausf H drive sprocket and at least one roadwheel has the cast hubcap which may have been introduced into production in October 1943. Note the single Bosch headlight and the non-standard Notek light an the hull front at the centre.

March 1944. No vehicles were completed in this month.

April 1944. Production commenced at Deutsche Eisenwerke with Fgst Nr.310371. The company initially reported that fourteen vehicles were completed in this month and accepted by the HWA but later revised this total to twenty. Curiously, the number is given as thirty and twelve in different official documents promulgated in May and June respectively.

The last Nashorn parts held at Alkett were shipped to Teplitz-Schönau. The Nachrichtenblatt der Panzertruppen for April advised that two Nashorn in each Stabs-Kompanie and one vehicle in each company headquarters were to be equipped with Fu 8 command radios in addition to the Fu 5. All other Nashorn were to be fitted with the Fu 5 model (1) which presumably replaced the Funksprechgerät f in production.

May 1944. The official production schedule dated 4 May projected that Deutsche Eisenwerke would complete thirty Nashorns in both April and May with the last forty vehicles of the series assembled in June, giving a total of 100. In the event twenty-four vehicles were completed in this month.

Ostketten, widened tracks designed to deal with soft terrain, were delivered to units serving on the Eastern Front.

June 1944. Just six vehicles were completed in this month, far short of the planned forty.

July 1944. A production schedule dated 14 July stated that Deutsche Eisenwerke was expected to complete twelve Nashorns in April, twenty-four in May, five in June and thirty in July, with a further thirty in August and twenty-nine in September. Despite these optimistic projections only three vehicles were built in this month.

August 1944. Thirty-one vehicles were built.

September 1944. Twelve vehicles were built.

October 1944. Seven vehicles were built.

November 1944. Five vehicles were built.

December 1944. No vehicles were built.

January 1945. A report dated 30 January stated that Stahlindustrie was expected to assemble nine Nashorn in this month with the last two vehicles of the series to be completed in February. In fact, all twelve were completed in January.

February 1945. Three vehicles were built.

March 1945. Just one vehicle, presumably Fgst Nr.310500, was completed in what would be the final month of production. Discussions to fit surplus 8.8cm Pak 43/1 guns to 180 Hummel chassis held at Duisburg eventually came to nothing although this proposal had been under consideration since at least mid-February.

During this month OKH reported that ninety-eight Nashorns were with operational units on all Fronts while seventy-three of those were considered combat ready.

Notes

1. In accordance with KstN 1148b and KAN 1155b, both dated 1 November1943.

Photographed in the East, this Nashorn is fitted with the extended track links known as Ostketten which would date this image to the late spring or early summer of 1944. Note the position of the U-shaped bracket which held the towing cables in place on the hull glacis. The gun travel lock seems to be missing the upper portion and the supporting brace is fixed in place with wire. This vehicle, and the recovery Sdkfz 9 halftrack in the background, are from schwere Panzerjager-Abteilung 88 and the battalion's unit insignia is just visible on the lower section of the

One of the twelve late-production Nashorn self-propelled guns assigned to 1.Kompanie, schwere Panzerjager-Abteilung 93 in December 1944. Note the bullet splash guard (1) added to the driver's visor. Note also that the upper portion (2) of the gun travel lock remains in place when the supporting brace (3) is lowered in preparation for firing. At least one detachable snow/ice cleat (4) is attached to the tracks which are the later versions with hollow guide horns (5).

2
3
1
4
5

SCHWERE PANZERJÄGER-ABTEILUNG, 1943-1945

All units of the German Army were organised according to orders and instructions issued by Oberkommando des Heeres (OKH), the high command of the army. These were accompanied by detailed charts referred to as Kriegsstärkenachweisung (KstN) which gave the authorised strength and composition of a unit listing the exact number of personnel, type of vehicle and weapons. Each KstN was individually numbered and dated, the latter being significant as many numbers were retained when an establishment was upgraded. They were complemented by identically numbered Kriegsausrüstungsnachweis (KAN) which listed smaller items such as uniforms and tools. KstN and KAN were not issued on a regular basis, as general orders were, but whenever a unit was formed or an organisational change was required. It was initially intended that companies equipped with the 8.8cm self-propelled gun should be attached to the anti-tank battalions of select Panzer divisions and the early establishments reflected this. The diagrams shown below and on the following pages describe the evolution of these battalions from semi-independent units to fully-formed battalions. At the risk of causing confusion I have alternated between the use of Hornisse and Nashorn, using the name that was correct at the time the orders were promulgated.

An OKH order dated 26 January 1943 directed that 4.Kompanie, Panzerjager-Abteilung 41 of 1.Panzer-Division be reorganised and equipped in accordance with KstN 1148a Ausf A dated 1 December 1942 with ten 8.8cm self-propelled guns. KAN 1148a also dated 1 December 1942 authorised one vehicle in each company to be fitted with an Fu 8 command radio and all vehicles to have Funksprechgerät f radio sets. The company was to be fully operational by 15 February 1943.

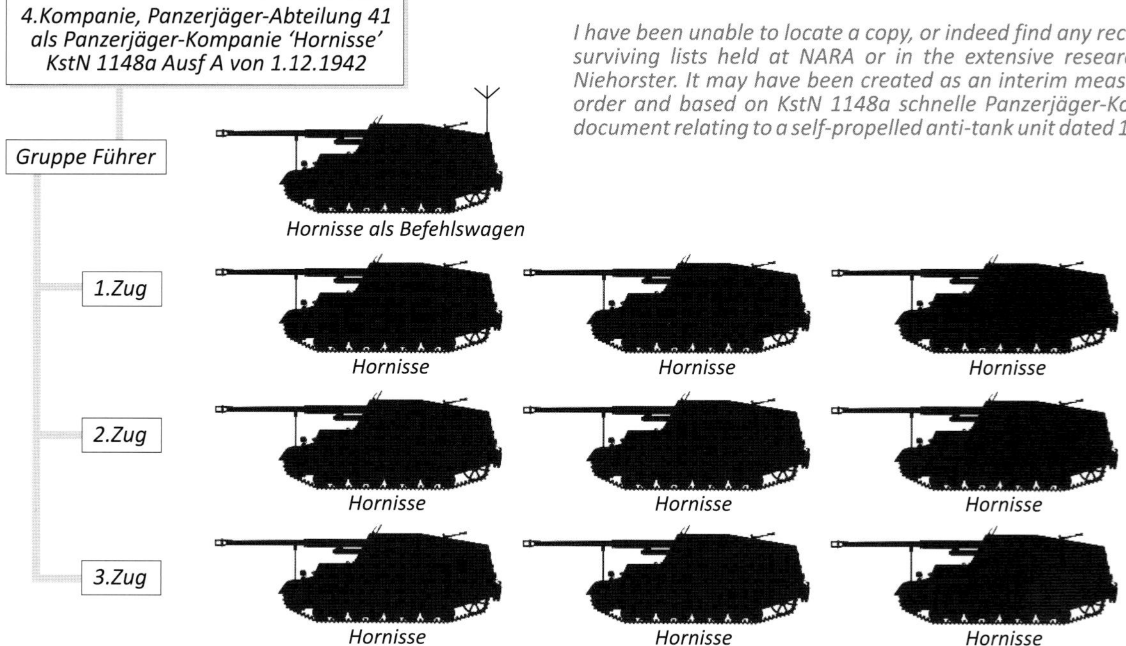

I have been unable to locate a copy, or indeed find any record, of this KstN on the surviving lists held at NARA or in the extensive research of both Tessin and Niehorster. It may have been created as an interim measure specifically for this order and based on KstN 1148a schnelle Panzerjäger-Kompanie (Sfl), the only document relating to a self-propelled anti-tank unit dated 1.12.1942.

An OKH order dated 6 February 1943 directed that 4.Kompanie, Panzerjäger-Abteilung 42 from 7.Panzer-Division be reorganised and equipped in accordance with KstN 1148b dated 30 January 1943 with ten 8.8cm Hornisse self-propelled guns. This was followed by an order of 10 February 1943 to form a third company based on 4.Kompanie, Panzerjäger-Abteilung 61 of 11.Panzer-Division using the same establishment.

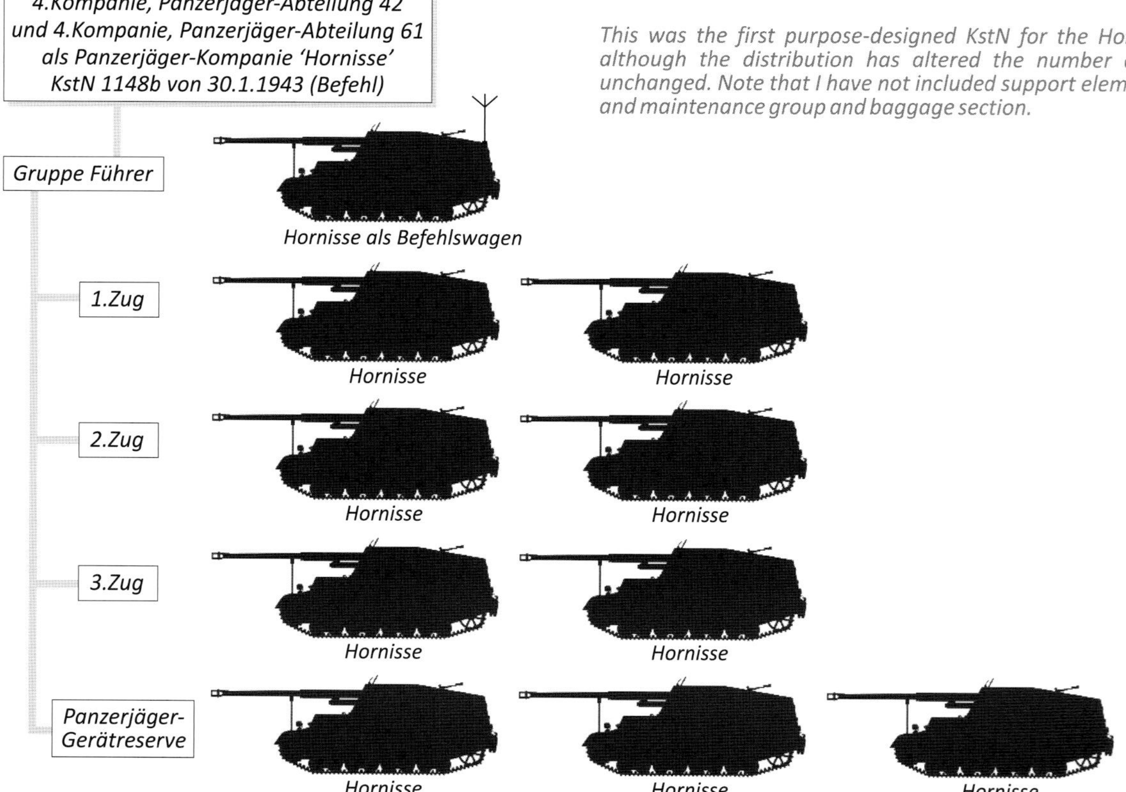

This was the first purpose-designed KstN for the Hornisse-Kompanien and although the distribution has altered the number of combat vehicles is unchanged. Note that I have not included support elements such as the repair and maintenance group and baggage section.

......continued on following page

By mid-February it had been decided that the three companies would not return to their original parent formations but would instead be assigned as follows:

Panzerjäger-Abteilung 37, 1.Panzer-Division

- *Stab* – KstN 1106 von 11.1.1941
- *1.Kompanie* – Self-propelled 7.62cm Pak KstN 1140 von 6.1.1942
- *2.Kompanie* – Towed 5cm Pak KstN 1142 von 1.11.1941
- *3.Kompanie* – 10 x Hornisse from 4.Kompanie, Panzerjäger-Abteilung 41
- *4.Kompanie* – Transferred to Panzergrenadier-Regiment 1 as 9(Flak).Kompanie.

Panzerjäger-Abteilung 41, 6.Panzer-Division

- *Stab* – KstN 1106 von 11.1.1941
- *1.Kompanie* – Self-propelled 7.62cm Pak KstN 1148d von 1.6.1943
- *2.Kompanie* – Towed 5cm Pak KstN 1142 von 1.11.1941
- *3.Kompanie* – Flak Kompanie KstN 192 von 11.1.1941
- *4.Kompanie*

10 x Hornisse from 4.Kompanie, Panzerjäger-Abteilung 42

Panzerjäger-Abteilung 42, 7.Panzer-Division

- *Stab* – KstN 1106 von 11.1.1941
- *1.Kompanie* – Towed 7.5cm Pak KstN 1140 von 1.4.1943
- *2.Kompanie* – Towed 5cm Pak KstN 1142 von 1.11.1941
- *3.Kompanie* – Flak Kompanie KstN 192 von 11.1.1941
- *4.Kompanie*

10 x Hornisse from 4.Kompanie, Panzerjäger-Abteilung 61

On 22 March 1943, before these changes could be implemented, OKH ordered that the three companies would be consolidated into a single battalion. An order of 24 March directed that the headquarters should be organised using KstN 1106a dated 1.11.1941 as a temporary guide (als Anhalt) but a further order of 25 March called for a headquarters and headquarters company to be formed according to KstN 1106b and KstN 1155b, both dated 30 March 1943, respectively. The new battalion, referred to as schwere Panzerjäger-Abteilung 560 (Hornisse), was to be tactically subordinated to 1.Panzer-Division. The revision of KstN 1155b of 30 March 1943 that was issued on 1 May 1943 seems to have affected personnel numbers in the support elements only.

Stab, Panzerjäger-Abteilung 'Hornisse'
KstN 1106b von 30.3.1943

- *Gefechtsstab* – Battalion Headquarters: 3 x Motorcycles with sidecar; 3 x Kfz Light Cars; 2 x Kfz 15 Medium Cars
- *Gepäcktross* – Battalion supply: 1 x 1.5-ton Lorry

The initial order mentions 'KstN und KAN 1106a (als Anhalt) vom 1.11.41 bis zum 15.4.1943 aufzustellen'. It appears that this organisation was to be used as a guide and that further clarification would follow in April. An OKH order dated 14.4.1943 stated 'KstN und KAN 1106b vom 30.3.1943, KstN und KAN 1155b vom 30.3.43 bis zum aufzustellen', meaning that this was also to be clarified at some future date.

Stabs-Kompanie, Panzerjäger-Abteilung 'Hornisse'
KstN 1155b von 30.3.1943

Gruppe Führer
Company Headquarters
2 x Motorcycles
1 x Kfz 15 Medium Car
3 x Hornisse

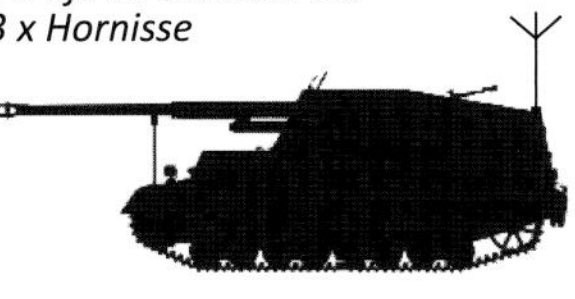

As the KAN is not available I have assumed here that all three vehicles were Hornisse als Befehlswagen based on a May 1943 combat report.

Gefechtstross
Company Supply Train
1 x Kfz 1 Light Car
3 x Lorry, 3-ton
2 x Lorry, 4.5-ton

Gepäcktross
Baggage Train
1 x Lorry, 3-ton

Verflegungstross
Ration Supply Train
1 x Light Car (Civilian)
2 x Lorry, 3-ton

Nachrichten-Staffel
Signals Detachment
2 x Motorcycles
1 x Kfz 15 Medium Car
1 x Kfz 2 Radio Car
1 x Kfz 17 Van

Pionier-Zug
Engineer Platoon
2 x Motorcycle
1 x Kfz Light Car
1 x Lorry, 3-ton

Fliegerabwehr-Zug
Anti-Aircraft Platoon
3 x Motorcycles
1 x Kfz 15 Medium Car

- *Flak-Trupp* – Anti-Aircraft Troop: 1 x Sdkfz 7, 2cm Flak; 1 x Sdkfz 7
- *Flak-Trupp* – Anti-Aircraft Troop: 1 x Sdkfz 7, 2cm Flak; 1 x Sdkfz 7
- *Flak-Trupp* – Anti-Aircraft Troop: 1 x Sdkfz 7, 2cm Flak; 1 x Sdkfz 7

Instandsetzung-Staffel
Maintenance and Repair Detachment
2 x Kfz Light Cars

- *Instandsetzung-Gruppe* – Maintenance Section: 1 x Kfz 40/2 Car; 4 x Lorry, 3-ton; 1 x Kfz 100 HebelKraft 3-ton Crane; 1 x Sdkfz 10 and trailer
- *Berge Gruppe* – Recovery Section: 1 x Kfz 1 Light Car; 3 x Sdkfz 9
- *Ersatzteil-Gruppe* – Spare Parts Section: 1 x Kfz 1 Light Car; 5 x Lorry, 3-ton
- *Gruppe für Waffen und Nachrichten Gerätinstandsetzung* – Armourer Section: 1 x Kfz 1 Light Car; 1 x Lorry, 3-ton

Staffel für Verwaltung und Nachschub
Battalion Supply Train
2 x Motorcycles
1 x Kfz Light Car
1 x Light Car (Civilian)
1 x Lorry, 3-ton
18 x Lorry, 4.5-ton

Sanitäts-Trupp
Medical Section
1 x Motorcycle with Sidecar
1 x Lorry, 3-ton
1 x Kfz 31 Ambulance
1 x Sdkfz 251/8

......continued on following page

The three companies that were preparing to join the Panzer divisions, 3.Kompanie, Panzerjäger-Abteilung 37, 4.Kompanie, Panzerjäger-Abteilung 41 and 4.Kompanie, Panzerjäger-Abteilung 42, were to be renamed as 1, 2 and 3.Kompanie respectively of schwere Panzerjäger-Abteilung 560. The headquarters units were to join 1.Panzer-Division, which was at that time being rebuilt in France, as soon as their organisation was complete. The Hornisse companies, which were to report directly to Oberbefehlshaber West, were expected to follow at some time after 29 March 1943. As late as 28 April it was noted in a status report that the companies were still organised under the 10-gun establishment and on 1 May 1943 an OKH order was issued that called for the companies of schwere Panzerjäger-Abteilung 560 to be reorganised according to KstN 1148b dated 1 April 1943 which authorised fourteen Hornisse for each company.

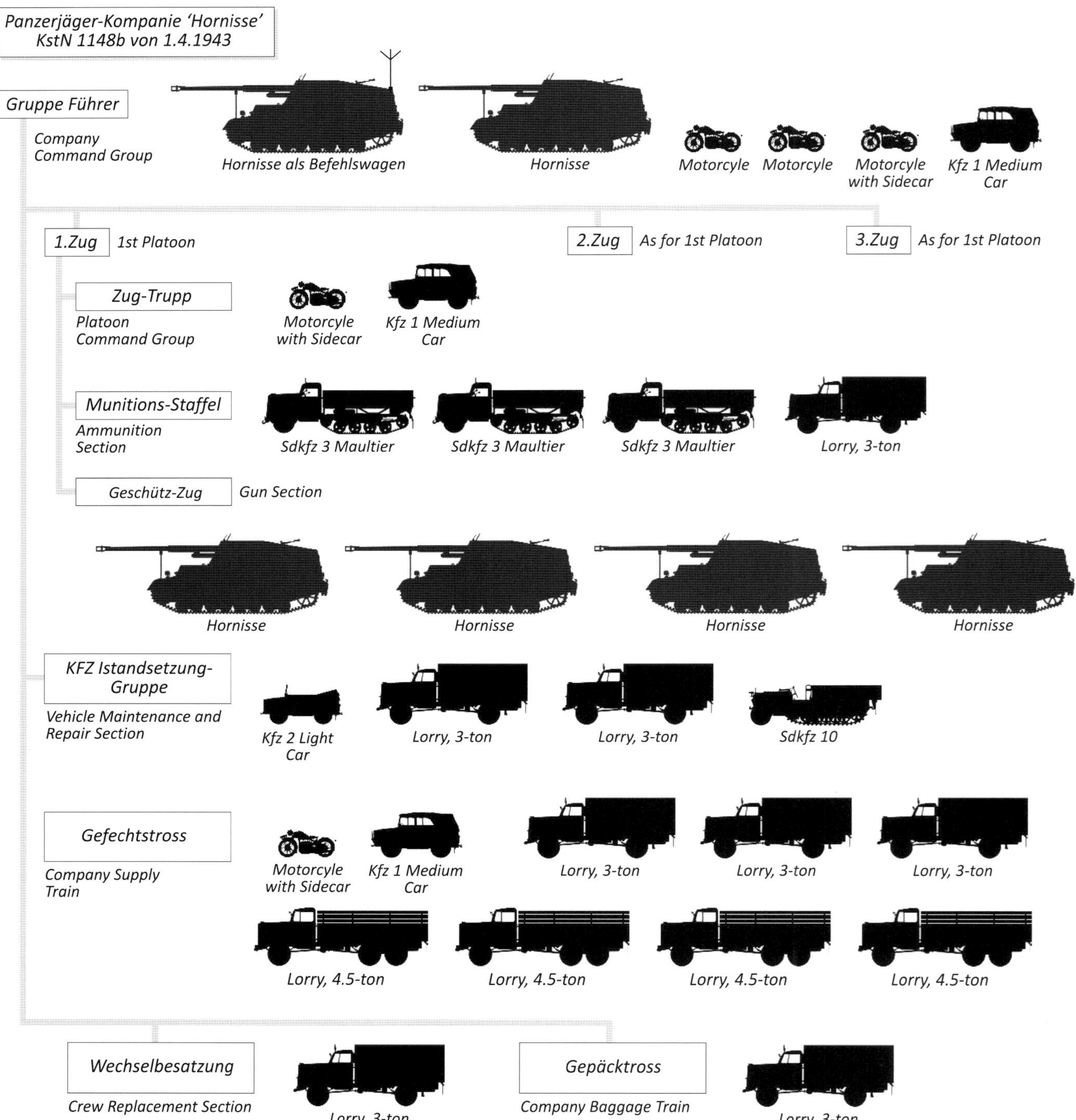

On 14 April 1943 OKH ordered that the units that had been created for Panzerjäger-Abteilung Stalingrad be used to form schwere Panzerjäger-Abteilung 655. The battalion headquarters and headquarters company were to be organised according to KstN 1106b and KstN 1155b respectively, both dated 30 March 1943. The three Hornisse companies were to be built using KstN 1148b of 1 April 1943. On 15 April 1943 the three Panzerjäger-Kompanien 521, 611 and 670 were renamed asd 1, 2 and 3.Kompanie, scwhere Panzerjäger-Abteilung 655.

Stab, Panzerjäger-Abteilung 'Hornisse'
KstN 1106b von 30.3.1943

Stabs-Kompanie, Panzerjäger-Abteilung 'Hornisse'
KstN 1155b von 30.3.1943

Panzerjäger-Kompanie 521
als Panzerjäger-Kompanie 'Hornisse'
KstN 1148b von 1.4.1943

Panzerjäger-Kompanie 611
als Panzerjäger-Kompanie 'Hornisse'
KstN 1148b von 1.4.1943

Panzerjäger-Kompanie 670
als Panzerjäger-Kompanie 'Hornisse'
KstN 1148b von 1.4.1943

......continued on following page

An OKH order of 24 April 1943 directed Panzerjäger-Ersatz- und Ausbildungs-Abteilung 9, a training and replacement unit, at Büdingen in western Germany to make available personnel for a schwere Panzerjäger-Abteilung 525 that would be organised according to KstN 1106b and KstN 1155b, both dated 30 March 1943, and KstN 1148b of 1.4.43. The following day an order was issued that stated Panzerjäger-Abteilung 525 was to be reorganised as a schwere Panzerjäger-Abteilung. And so the first three battalions were, eventually, organised identically and fully equipped with forty-five Hornisse by June 1943.

Stab, Panzerjäger-Abteilung 'Hornisse'
KstN 1106b von 30.3.1943

Stabs-Kompanie, Panzerjäger-Abteilung 'Hornisse'
KstN 1155b von 30.3.1943

Panzerjäger-Kompanie 'Hornisse'
KstN 1148b von 1.4.1943

Panzerjäger-Kompanie 'Hornisse'
KstN 1148b von 1.4.1943

Panzerjäger-Kompanie 'Hornisse'
KstN 1148b von 1.4.1943

On 1 August 1943 OKH ordered that Panzerjäger-Abteilung 93, which had been subordinated to 26.Panzer-Division, be renamed schwere Panzerjäger-Abteilung 93. The battalion headquarters was to be reorganised according to KstN 1106b of 30 March 1943, the headquarters company using KstN 1155b and the companies built with KstN 1148b, both dated 28 August 1943. Interestingly, a further OKH order dated 28 August gives a slightly different organisation which, in the interests of accuracy, I will quote here in the original German:

'II. Gliederung: Stab ...gem.KstN und KAN 1106b v.30.3.43. Stabskp...gem. KstN und KAN 115 b v.28.3.43.3 Kp... gem. KstN und KAN 1148 b v.28.8.43'.

The prefix 'gem' applied to each KstN signifies Gemishte, or mixed, but I am unable to explain its use here. These establishments may have been a temporary expedient as a number of the battalion's towed anti-tank guns were still on hand although they are not mentioned again and unfortunately no copies seem to have survived the war.

On 27 August 1943 OKH ordered that Panzerjäger-Ersatz- und Ausbildungs-Abteilung 43 begin the formation of schwere Panzerjäger-Abteilung 519. Once again, I will include a direct quote from the order:

'In der Gliederung: Stab gem.KstN und KAN 1106b vom 30.3.43, Stabs-Kp. gem.KstN und KAN 1155b vom 28.8.43, 3 Kp. gem.KstN und KAN 1148b vom 1.4.43.'

The reference to Gemischte, or mixed, formations is even more perplexing here as this battalion was not based on an existing anti-tank unit. Paragraph V of the order regarding KstN and KAN states: 'gehen gesondert durch Chef H Rüst und BdE/AHA/V zu' meaning that separate orders were to follow from the Chief of Army Armaments and the Commander of the Replacement Army. Nevertheless, it is tempting to speculate that units equipped with both the towed and self-propelled versions of the 8.8cm Pak 43 were at some time contemplated. On 17 September 1943 new orders were issued that called for the battalion headquarters to be reorganised according to KstN 1106b of 30 March 1943, the headquarters company using KstN 1155b and the companies built with KstN 1148b, both dated 28 August 1943.

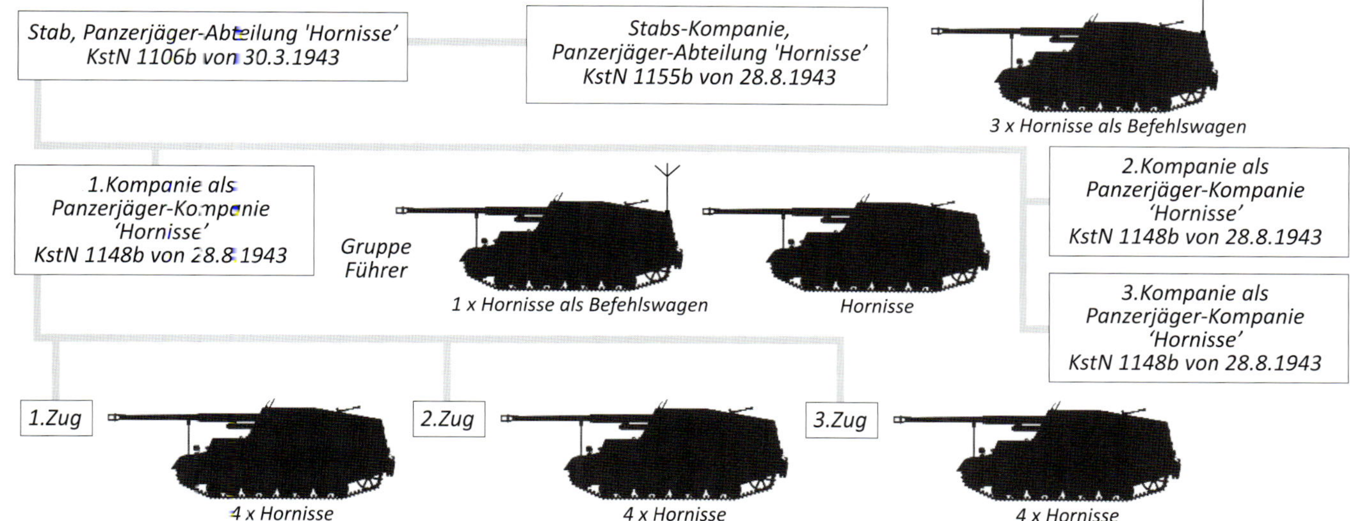

The changes to KstN 1155b and KstN 1148b saw a slight reduction in personnel and a number of transport elements transferred from the individual companies and handed to the battalion headquarters company. The number of Hornisse authorised for the Stabs-Kompanie and Panzerjäger-Kompanien, as can be seen here, was unchanged.

On 3 December 1943 the last of the true Hornisse battalions was created when OKH ordered that Panzerjäger-Abteilung 88, which had been subordinated to 13.Panzer-Division, be reorganised as a heavy anti-tank battalion and renamed schwere Panzerjäger-Abteilung 88. The battalion headquarters was to be reorganised according to KstN 1106b of 30 March 1943 and revised establishments for KstN 1155b and KstN 1148b, the headquarters company and Panzerjäger companies respectively, were dated 1 November 1943. Again, most changes affected the transport elements and this probably reflected the growing trend to concentrate support units at battalion or regimental headquarters which would culminate in the freie Gliederung organisations. The battalion was fully equipped by late January 1944.

In August 1944 the last crews of schwere Panzerjäger-Abteilung 519 had been transferred to Germany and the battalion's surviving vehicles were handed over to schwere Heeres-Panzerjäger-Abteilung 664, a unit equipped with towed 8.8cm Pak 43 guns. The latter were concentrated in 1.Kompanie and 3.Kompanie while the self-propelled guns formed a new 2.Kompanie. The organisation shown here is slightly conjectural but highly likely.

Stab, Panzerjäger-Abteilung (Mot)
KstN 1106 von 1.11.1941

1.Kompanie als
schwere Panzerjäger-Kompanie
(zu 12 Geschützen 8,8cm Pak 43) (mot Z)
KstN 1146 von 15.5.1943

2.Kompanie als
Panzerjäger-Kompanie 'Hornisse'
KstN 1148b von 28.8.1943

3.Kompanie als
schwere Panzerjäger-Kompanie
(zu 12 Geschützen 8,8cm Pak 43) (mot Z)
KstN 1146 von 15.5.1943

Dragon Models Ltd
B1-10/F., 603-609 Castle Peak Rd,
Kong Nam Industrial Building,
Tsuen Wan, N. T., Hong Kong
www.dragon-models.com

Tamiya Inc
Shizuoka City, Japan
www.tamiya.com

Hobby Fan/ AFV Club
6F., No.183, Sec. 1, Datong Rd, Xizhi City,
Taipei County 221, Taiwan
www.hobbyfan.com

Box Studio
Offering paper models. Unfortunately the web site is in Japanese only (although there is supposed to be an English version).
www.boxstudio.main.jp/

Zvezda
www.zvezda.org.ru

Roco Minitanks
Selected models available through Herpa Miniaturmodelle
www.herpa.de/en
or on-line hobby stores and auction sites

Planet Models
Available from CMK Kits
Mezilesí 718/78
19300, Prague 9
Czech Republic
www. cmkkits.com/en

Bolt Action
Available through Warlord Games
www.warlord.games.com
There are also many international outlets

Border Model
Xizhang Industrial Park, Huishan District,
Wuxi City, P.R. China
www.bordermodel.com

Voyager
Offering highly detail, comprehensive photo-etched brass detail sets.
Room 501, No.411 4th Village,
SPC Jinshan District, Shanghai 200540
P.R.China
www.voyagermodel.com

Griffon Model
A manufacturer of high-quality aftermarket details in resin and photo-etched brass.
Suite 501, Bldg 01, 418 Middle Longpan Rd,
Nanjing, P.R. China
www.griffonmodel.com

Aber
This company produces an enormous range of addtional detail in 1/72, 1/48 and 1/35 scale.
ul. Jalowcowa 15, 40-750 Katowice, Poland
www.aber.net.pl

E.T. Model
Highly-detailed brass and resin upgrades at a very reasonable price.
www.etmodeller.com

Friulmodel
Specialising in highly-accurate, workable scale track in white metal
H 8142. Urhida, Nefelejcs u. 2., Hungary
www.friulmodel.hu

Modelkasten
Workable scale track links produced in injection moulded plastic.
Chiyoda-ku Kanda, Nishiki-Cho 1-7, Tokyo, Japan
www.modelkasten.com
Very difficult to navigate but worthwhile.

Eduard Model Accessories
Regular readers will be familiar with this manufacturer of quality upgrade sets.
Mirova 170, 435 21 Obrnice,
Czech Republic
www.eduard.com

Hauler
A manufacturer of aftermarket details in resin, photo-etched brass and aluminium.
Jan Sobotka,
Moravská 38, 620 00 Brno,
Czech Republic
www.hauler.cz

Master Club
This company produces some of the best workable metal tarck links on the market. It appears that this firm is closely associated with Armour35, a Russian mail-order firm.
www.masterclub.ru

Kaizen
Like many Chinese companies this is hard to pin down to an exact location. The tracks are available from MS Models Japan, a firm for which I can personally vouch.
www.msmodelswebshop.jp

Royal Model
This company produces a large range of additional details, upgrade sets and figures.
Via E. Montale, 19-95030 Pedara, Italy
www.royalmodel.com

Sturmpanzer
Copies of original documents including tables of organisation, production statistics and allocations, all at extremely reasonable prices. I would highly recommend this service.
www.sturmpanzer.com

In writing this book I referred extensively to the Panzer Tracts series of booklets created by the late Thomas Jentz and Hilary Doyle, particularly *Panzer Tracts 7-3: Panzerjaeger (7,5cm PaK 40/4 to 8.8cm Waffentraeger) Development and Employment from 1939 to 1945* and *Panzer Tracts 23: Panzer Production From 1939 to 1945*. I also referred to Thomas Anderson's *The History of the Panzerjäger Volume 2: From Stalingrad to Berlin, 1943-1945* and Janusz Lewoch's *Nashorn* which contains some very useful references to the vehicle exhaust modifications. The information on unit allocations was taken, for the most part, from original Zufürungliste, or supply lists, compiled by the Heereszeugamt between May 1943 and April 1945. Similarly, the official OKH orders which concerned the formation of the original four battalions were consulted in researching the unit histories. I also relied heavily on the work of Martin Block and the late Ron Klages whose research on vehicle allocations, including those which actually arrived at their destination, would fill many volumes. I would also like to thank Lisa Hooson and Stephen Chumbley, my editors at Pen & Sword, and the very talented modellers who allowed me to include their work. As always, I am indebted to Karl Berne, Valeri Polokov and J.Howard Parker for their invaluable assistance with the photographs and period insignia.

A Nashorn of schwere Panzerjäger-Abteilung 519 named Puma preparing to fire on enemy positions during the fighting for Vitebsk in late 1943. Note the spare roadwheel behind the vehicle jack, the open driver's hatch and the Scherenfernrohr binocular periscope.